SpringerBriefs in Mathematics

SpringerBriefs in Mathematics showcases expositions in all areas of mathematics and applied mathematics. Manuscripts presenting new results or a single new result in a classical field, new field, or an emerging topic, applications, or bridges between new results and already published works, are encouraged. The series is intended for mathematicians and applied mathematicians.

More information about this series at http://www.springer.com/series/10030

SBMAC SpringerBriefs

The **SBMAC SpringerBriefs** series publishes relevant contributions in the fields of applied and computational mathematics, mathematics, scientific computing, and related areas. Featuring compact volumes of 50 to 125 pages, the series covers a range of content from professional to academic.

The Sociedade Brasileira de Matemática Aplicada e Computacional (Brazilian Society of Computational and Applied Mathematics, SBMAC) is a professional association focused on computational and industrial applied mathematics. The society is active in furthering the development of mathematics and its applications in scientific, technological, and industrial fields. The SBMAC has helped to develop the applications of mathematics in science, technology, and industry, to encourage the development and implementation of effective methods and mathematical techniques for the benefit of science and technology, and to promote the exchange of ideas and information between the diverse areas of application.

http://www.sbmac.org.br/

Antonio André Novotny • Jan Sokołowski

An Introduction to the Topological Derivative Method

 Springer

Antonio André Novotny
Coordenação de Métodos
Matemáticos e Computacionais
Laboratório Nacional de Computação
Científica LNCC/MCTIC
Petrópolis, Rio de Janeiro, Brazil

Jan Sokołowski
Institut Élie Cartan de Nancy, UMR 7502
Université de Lorraine, CNRS
Vandoeuvre-Lès-Nancy, France

Systems Research Institute
Polish Academy of Sciences
Warsaw, Poland

Universidade Federal da Paraiba, Centro de
informatica
João Pessoa, PB, Brazil

ISSN 2191-8198   ISSN 2191-8201   (electronic)
SpringerBriefs in Mathematics
ISBN 978-3-030-36914-9   ISBN 978-3-030-36915-6   (eBook)
https://doi.org/10.1007/978-3-030-36915-6

Mathematics Subject Classification: 49Q12, 35C20, 35A15, 74B05, 49J20, 35Q93

This Springer imprint is published by the registered company Springer Nature Switzerland AG.
The registered company address is: Gewerbestrasse 11, 6330 Cham, Switzerland

This book is dedicated to
Bożena and Daniel & Vanessinha

Preface

Mathematical analysis and numerical solutions of problems with unknown shapes is a challenging and rich research field in the modern theory of calculus of variations, partial differential equations, differential geometry as well as in numerical analysis. In this book, the topological derivative method is introduced. This new concept in shape optimization has applications in many different fields such as topology optimization, inverse problems, imaging processing, multi-scale material design, and mechanical modeling including damage and fracture evolution phenomena.

The present work is complementary to the book by A.A. Novotny and J. Sokołowski, *Topological Derivatives in Shape Optimization, Interaction of Mechanics and Mathematics Series*, Springer-Verlag, Berlin, Heidelberg, 2013. In fact, the concept of topological derivative is presented through some selected examples in a simpler and more pedagogical manner by using a direct approach based on calculus of variations combined with compound asymptotic analysis. In addition, the topological derivative is used in numerical method of shape optimization, including applications in the context of compliance structural topology optimization and topology design of compliant mechanisms. Finally, in order to fix the ideas, some exercises are proposed at the end of each chapter. Therefore, this monograph can be adopted as a textbook in graduate and introductory courses on the subject. In particular, it is oriented to researchers and students in applied mathematics and computational mechanics interested in the mathematical aspects of the topological asymptotic analysis as well as in applications of the topological derivative method in computational mechanics.

This book is a result of more than 10 years of scientific collaboration between André Novotny and Jan Sokołowski, which has been supported by IECN in France and by CNPq, CAPES, FAPERJ, and LNCC in Brazil.

Petrópolis, Brazil
Vandoeuvre-Lès-Nancy, France
Warsaw, Poland
João Pessoa, Brazil

Antonio André Novotny
Jan Sokołowski

Contents

Chapter 1
Introduction

The topological derivative is defined as the first term of the asymptotic expansion of a given shape functional with respect to a small parameter that measures the size of singular domain perturbations, such as holes, inclusions, source-terms, and cracks [75]. This relatively new concept has applications in many different fields such as shape and topology optimization, inverse problems, imaging processing, multi-scale material design, and mechanical modeling including damage and fracture evolution phenomena. For an account on the theoretical development and applications of the topological derivative method, see the series of review papers [79–81] and references therein.

The topological derivative method has been specifically designed to deal with topology optimization. It has been introduced by Sokołowski and Żochowski in 1999 through the fundamental paper [88] to fill a gap in the existing literature at that time. Actually, the idea was to give a precise (mathematical) answer to the following important question: What happens when a hole is nucleated? The answer to this question is not trivial at all, since singularities may appear once a hole is nucleated. Therefore, in order to deal with this problem, asymptotic analysis in singularly perturbed geometrical domains is needed. In this book, the topological derivative method is presented through some selected examples in a simple and pedagogical manner by using a direct approach based on calculus of variations combined with *matched* [54] and *compound* [67] asymptotic analysis of solution to boundary value problems. In addition, the topological derivative is used in numerical method of shape optimization including applications in the context of compliance structural topology optimization and topology design of compliant mechanisms. Finally, some exercises are proposed at the end of each chapter for the readers' convenience.

This chapter is organized as follows. In Sect. 1.1 the topological derivative concept is introduced and five simple examples are presented in order to fix the ideas. In Sect. 1.2 the adjoint sensitivity method is presented through the Lagrangian formalism and an example in the context of control theory with PDE constraint

A. A. Novotny, J. Sokołowski, *An Introduction to the Topological Derivative Method*, SpringerBriefs in Mathematics, https://doi.org/10.1007/978-3-030-36915-6_1

is fully developed. Finally, the content of the book is described in detail through Sect. 1.3.

1.1 The Topological Derivative Concept

Let us consider an open and bounded domain $\Omega \subset \mathbb{R}^d$, with $d \geq 2$, which is subject to a non-smooth perturbation confined in a small region $\omega_\varepsilon(\widehat{x}) = \widehat{x} + \varepsilon \omega$ of size ε, such that $\overline{\omega_\varepsilon} \subset \Omega$, where $\widehat{x}$ is an arbitrary point of Ω and ω represents a fixed domain in $\mathbb{R}^d$. See sketch in Fig. 1.1. We introduce a *characteristic function* $x \mapsto \chi(x)$, $x \in \mathbb{R}^d$, associated with the unperturbed domain, namely $\chi := \mathbb{1}_\Omega$, such that

$$|\Omega| = \int_{\mathbb{R}^d} \chi, \qquad (1.1)$$

where $|\Omega|$ is the *Lebesgue measure* of Ω. Then, we define a characteristic function associated with the topologically perturbed domain of the form $x \mapsto \chi_\varepsilon(\widehat{x}; x)$, $x \in \mathbb{R}^d$. In the case of a perforation, for example, $\chi_\varepsilon(\widehat{x}) := \mathbb{1}_\Omega - \mathbb{1}_{\omega_\varepsilon(\widehat{x})}$ and the perturbed domain is obtained as $\Omega_\varepsilon = \Omega \setminus \overline{\omega_\varepsilon}$. Finally, we assume that a given shape functional $\psi(\chi_\varepsilon(\widehat{x}))$, associated with the topologically perturbed domain, admits a *topological asymptotic expansion* of the form

$$\psi(\chi_\varepsilon(\widehat{x})) = \psi(\chi) + f(\varepsilon)\mathscr{T}(\widehat{x}) + \mathscr{R}(\varepsilon), \qquad (1.2)$$

where $\psi(\chi)$ is the shape functional associated with the reference (unperturbed) domain, $f(\varepsilon)$ is a positive *first order correction function*, which decreases monotonically such that $f(\varepsilon) \to 0$ with $\varepsilon \to 0$, and $\mathscr{R}(\varepsilon)$ is the *remainder term*, that is, $\mathscr{R}(\varepsilon)/f(\varepsilon) \to 0$ with $\varepsilon \to 0$. The function $\widehat{x} \mapsto \mathscr{T}(\widehat{x})$ is recognized as the *topological derivative* of ψ at $\widehat{x}$. Therefore, the product $f(\varepsilon)\mathscr{T}(\widehat{x})$ represents a first order correction over $\psi(\chi)$ to approximate $\psi(\chi_\varepsilon(\widehat{x}))$. In addition, after rearranging (1.2), we have

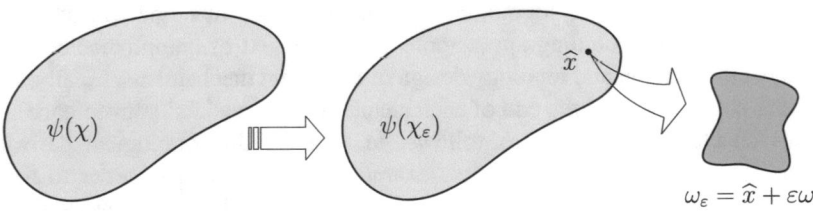

Fig. 1.1 The topological derivative concept

$$\frac{\psi(\chi_\varepsilon(\widehat{x})) - \psi(\chi)}{f(\varepsilon)} = \mathscr{T}(\widehat{x}) + \frac{\mathscr{R}(\varepsilon)}{f(\varepsilon)}. \tag{1.3}$$

The limit passage $\varepsilon \to 0$ in the above expression leads to the general definition for the *topological derivative*, namely

$$\mathscr{T}(\widehat{x}) := \lim_{\varepsilon \to 0} \frac{\psi(\chi_\varepsilon(\widehat{x})) - \psi(\chi)}{f(\varepsilon)}. \tag{1.4}$$

Assuming that the functional $\psi(\chi_\varepsilon(\widehat{x}))$ admits the topological asymptotic expansion (1.2), the applicability of this expansion depends on the procedure of evaluation of the unknown function $\widehat{x} \mapsto \mathscr{T}(\widehat{x})$. In particular, we are looking for an appropriate form for the topological derivative which can be used in numerical method of shape/topology optimization, for instance. Therefore, we need some properties of the shape functional and its asymptotic expansion in order to apply a simple method for evaluation of the topological derivative, which are:

1. The shape functional $\varepsilon \mapsto j(\varepsilon) := \psi(\chi_\varepsilon(\widehat{x}))$ is continuous with respect to a *topological perturbation* at 0^+, i.e., $\lim_{\varepsilon \to 0^+} f(\varepsilon) = 0$.
2. The limit passage $\lim_{\varepsilon \to 0^+} \mathscr{R}(\varepsilon)/f(\varepsilon) = 0$ holds true.

Note that since we are dealing with singular domain perturbations, in general the limit in (1.4) cannot be trivially evaluated. It is the case of topological perturbations consisting of nucleation of holes, for instance, where the shape functionals $\psi(\chi)$ and $\psi(\chi_\varepsilon(\widehat{x}))$ are associated with topologically different domains (see Remark 1.2 below). Therefore, we need to develop $\psi(\chi_\varepsilon(\widehat{x}))$ asymptotically with respect to the small parameter ε and collect the leading terms of the resulting expansion. How to construct a topological asymptotic expansion of the form (1.4) is, in fact, the main concern of this monograph.

Remark 1.1 The notion of topological derivative extends the conventional definition of derivative [34, 35], allowing to deal with functionals depending on a *geometrical domain* subjected to singular topology changes. According to (1.4), the analogy between the topological *derivative* and the corresponding expressions for a conventional derivative is to be noted.

Remark 1.2 We say *topological* derivative because we are dealing with topological changes in a geometrical domain given by, e.g., nucleation of holes. In fact, the *Euler-Poincaré characteristic* of any oriented surface Ω is given by the quantity

$$C(\Omega) = V - E + F, \tag{1.5}$$

where V, E, and F are respectively the numbers of vertices, edges, and faces of a given polyhedron produced by an arbitrary triangularization of Ω. In particular, if two distinct surfaces Ω_1 and Ω_2 have the same Euler-Poincaré characteristic, namely $C(\Omega_1) = C(\Omega_2)$, then they are topologically equivalents. Let us now suppose that we remove one single triangle from Ω_2, then Ω_1 and Ω_2 have the

Fig. 1.2 An example of the
Euler-Poincaré characteristic

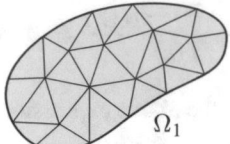
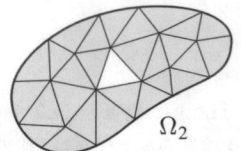

same numbers of vertices and edges, but Ω_2 has one face less than Ω_1, so that $C(\Omega_2) = C(\Omega_1) - 1$. See sketch in Fig. 1.2. Therefore, according to the Euler-Poincaré characteristic, after creating a hole in Ω_2 by removing a triangle, its topology actually changes.

In order to fix these ideas let us present five (very) simple examples. The first one concerns the topological derivative of the volume of a given geometrical domain. The second and third examples deal with singular and regular domain perturbations, respectively. The fourth example shows that the topological derivative obeys the basic rules of differential calculus. Finally, the last example deals with the topological derivative of the energy shape functional associated with a second order ordinary differential equation into one spatial dimension.

Example 1.1 Let us consider a very simple functional given by the area of the domain $\Omega \subset \mathbb{R}^2$, that is

$$\psi(\chi) := |\Omega| = \int_\Omega 1, \tag{1.6}$$

with Ω subject to the class of topological perturbations produced by the nucleation of circular holes, namely $\omega_\varepsilon = B_\varepsilon(\widehat{x}) := \{x \in \Omega : \|x - \widehat{x}\| < \varepsilon\}$, for $\widehat{x} \in \Omega$. The expansion with respect to ε can be trivially obtained as follows:

$$\psi(\chi_\varepsilon(\widehat{x})) = |\Omega_\varepsilon(\widehat{x})| = \int_\Omega 1 - \int_{B_\varepsilon} 1 = \psi(\chi) - \pi\varepsilon^2. \tag{1.7}$$

Therefore, function $f(\varepsilon)$ and the topological derivative $\mathscr{T}(\widehat{x})$ are immediately identified as

$$f(\varepsilon) = \pi\varepsilon^2 \quad \text{and} \quad \mathscr{T}(\widehat{x}) = -1 \quad \forall \widehat{x} \in \Omega. \tag{1.8}$$

In this particular case $\mathscr{T}(\widehat{x})$ is independent of $\widehat{x}$, and the rightmost term of the topological asymptotic expansion is equal to zero.

Example 1.2 We consider a shape functional of the form

$$\psi(\chi_\varepsilon(\widehat{x})) := \int_{\Omega_\varepsilon(\widehat{x})} g(x), \tag{1.9}$$

where $\chi_\varepsilon(\widehat{x}) = \mathbb{1}_\Omega - \mathbb{1}_{B_\varepsilon(\widehat{x})}$ and $\Omega_\varepsilon(\widehat{x}) = \Omega \setminus \overline{B_\varepsilon(\widehat{x})}$, with $B_\varepsilon(\widehat{x})$ used to denote a ball of radius ε and center at $\widehat{x} \in \Omega$. The function $g : \mathbb{R}^2 \mapsto \mathbb{R}$ is assumed to be

Lipschitz continuous in $B_\varepsilon(\widehat{x})$, i.e., $|g(x) - g(\widehat{x})| \leq C\|x - \widehat{x}\| \; \forall x \in B_\varepsilon(\widehat{x})$, where $C \geq 0$ is the Lipschitz constant. In the context of problems governed by partial differential equations, this property comes out from the interior *elliptic regularity* of solutions. Note that this is the case regarding singular domain perturbation (see Remark 1.2). Since $|B_\varepsilon| \to 0$ with $\varepsilon \to 0$, we have

$$\psi(\chi) := \int_\Omega g(x). \tag{1.10}$$

We are looking for an asymptotic expansion of the form (1.2), namely

$$\begin{aligned}
\psi(\chi_\varepsilon(\widehat{x})) &= \int_{\Omega_\varepsilon} g(x) + \int_{B_\varepsilon} g(x) - \int_{B_\varepsilon} g(x) \\
&= \int_\Omega g(x) - \int_{B_\varepsilon} g(x) + \int_{B_\varepsilon} g(\widehat{x}) - \int_{B_\varepsilon} g(\widehat{x}) \\
&= \psi(\chi) - \pi\varepsilon^2 g(\widehat{x}) + \mathscr{E}(\varepsilon).
\end{aligned} \tag{1.11}$$

The remainder $\mathscr{E}(\varepsilon)$ is defined as

$$\mathscr{E}(\varepsilon) = -\int_{B_\varepsilon} (g(x) - g(\widehat{x})), \tag{1.12}$$

which can be bounded as follows:

$$|\mathscr{E}(\varepsilon)| = \left| \int_{B_\varepsilon} (g(x) - g(\widehat{x})) \right| \leq \int_{B_\varepsilon} |g(x) - g(\widehat{x})| \leq C_1 \int_{B_\varepsilon} \|x - \widehat{x}\|, \tag{1.13}$$

since function g is assumed to be locally Lipschitz continuous. From a polar coordinate system (r, θ) centered at the point $\widehat{x} \in \Omega$, there is

$$\int_{B_\varepsilon} \|x - \widehat{x}\| = \int_0^{2\pi} \left(\int_0^\varepsilon (r)\, r dr \right) d\theta = \frac{2\pi}{3}\varepsilon^3, \tag{1.14}$$

where we have used the fact that $\|x - \widehat{x}\| = r$. Finally, by combining the last two results, the following estimate for the remainder $\mathscr{E}(\varepsilon)$ holds true:

$$|\mathscr{E}(\varepsilon)| \leq C_2\varepsilon^3, \tag{1.15}$$

with constant C_2 independent of the small parameter ε. Therefore, the term $-g(\widehat{x})$ is identified as the topological derivative of the shape functional ψ evaluated at the point $\widehat{x} \in \Omega$, namely

$$\mathscr{T}(\widehat{x}) = -g(\widehat{x}) \quad \forall \widehat{x} \in \Omega. \tag{1.16}$$

In addition, function $f(\varepsilon) = \pi\varepsilon^2$. Finally, the remainder $\mathscr{E}(\varepsilon)$ is of order $O(\varepsilon^3)$.

Example 1.3 Now, let us consider a shape functional defined as follows:

$$\psi(\chi_\varepsilon(\widehat{x})) := \int_\Omega g_\varepsilon(x), \tag{1.17}$$

where $\chi_\varepsilon(\widehat{x}) = \mathbb{1}_\Omega - (1 - \gamma)\mathbb{1}_{B_\varepsilon(\widehat{x})}$. The function $g_\varepsilon = \chi_\varepsilon g$ is defined as

$$g_\varepsilon(x) := \begin{cases} g(x) \text{ if } x \in \Omega \setminus \overline{B_\varepsilon(\widehat{x})}, \\ \gamma g(x) \text{ if } x \in B_\varepsilon(\widehat{x}), \end{cases} \tag{1.18}$$

where $\gamma \in \mathbb{R}$ is the contrast and $B_\varepsilon(\widehat{x})$ is a ball of radius ε and center at $\widehat{x} \in \Omega$. In addition, function $g : \mathbb{R}^2 \mapsto \mathbb{R}$ is assumed to be Lipschitz continuous in B_ε (see Example 1.2). Observe that this case corresponds to regular domain perturbation where the shape functional depends on characteristic function of small sets. Since $|B_\varepsilon| \to 0$ with $\varepsilon \to 0$, there is

$$\psi(\chi) := \int_\Omega g(x). \tag{1.19}$$

We are looking for an asymptotic expansion of the form (1.2), that is

$$\begin{aligned} \psi(\chi_\varepsilon(\widehat{x})) &= \int_{\Omega \setminus \overline{B_\varepsilon}} g(x) + \int_{B_\varepsilon} \gamma g(x) \\ &= \int_{\Omega \setminus \overline{B_\varepsilon}} g(x) + \int_{B_\varepsilon} \gamma g(x) \pm \int_{B_\varepsilon} g(x) \\ &= \int_\Omega g(x) - (1 - \gamma) \int_{B_\varepsilon} g(x) \pm (1 - \gamma) \int_{B_\varepsilon} g(\widehat{x}) \\ &= \psi(\chi) - \pi\varepsilon^2(1 - \gamma)g(\widehat{x}) + o(\varepsilon^2), \end{aligned} \tag{1.20}$$

where we have used the notation $0 = +(\cdot) = (\cdot) - (\cdot)$. From the above expansion, we can identify the term $-(1 - \gamma)g(\widehat{x})$ as the topological derivative of the shape functional ψ evaluated at the point $\widehat{x} \in \Omega$, namely

$$\mathscr{T}(\widehat{x}) = -(1 - \gamma)g(\widehat{x}) \quad \forall \widehat{x} \in \Omega, \tag{1.21}$$

with $f(\varepsilon) = \pi\varepsilon^2$. Note that the limit passage $\gamma \to 0$ leads to $\mathscr{T}(\widehat{x}) = -g(\widehat{x})$. It means that the former example can be seen as the singular limit of this one.

Example 1.4 We consider two functions g_1 and g_2 assumed to be Lipschitz continuous in B_ε. Let us return to the case regarding singularly perturbed geometrical domains of the form $\Omega_\varepsilon(\widehat{x}) = \Omega \setminus \overline{B_\varepsilon(\widehat{x})}$. According to Example 1.2, we have

$$\psi_i(\chi) := \int_\Omega g_i(x) \quad \Rightarrow \quad \mathscr{T}_i(\widehat{x}) = -g_i(\widehat{x}), \quad \text{for} \quad i = 1, 2. \tag{1.22}$$

Then, the topological derivative of the product between $\psi_1(\chi)$ and $\psi_2(\chi)$, namely

$$\psi(\chi):=\psi_1(\chi)\psi_2(\chi),\tag{1.23}$$

is given by

$$\begin{aligned}\mathscr{T}(\widehat{x}) &= \mathscr{T}_1(\widehat{x})\psi_2(\chi) + \mathscr{T}_2(\widehat{x})\psi_1(\chi)\\ &= -g_1(\widehat{x})\int_\Omega g_2(x) - g_2(\widehat{x})\int_\Omega g_1(x).\end{aligned}\tag{1.24}$$

Finally, the topological derivative of the quotient between $\psi_1(\chi)$ and $\psi_2(\chi)$, that is

$$\psi(\chi):=\frac{\psi_1(\chi)}{\psi_2(\chi)},\tag{1.25}$$

can be written as

$$\begin{aligned}\mathscr{T}(\widehat{x}) &= \frac{\mathscr{T}_1(\widehat{x})\psi_2(\chi) - \mathscr{T}_2(\widehat{x})\psi_1(\chi)}{\psi_2(\chi)^2}\\ &= \frac{g_2(\widehat{x})\int_\Omega g_1(x) - g_1(\widehat{x})\int_\Omega g_2(x)}{\psi_2(\chi)^2}.\end{aligned}\tag{1.26}$$

Example 1.5 In this last example we consider that the problem is governed by a second order ordinary differential equation. The associated energy shape functional is defined as

$$\psi_\Gamma(\chi_\varepsilon(\widehat{x})) := \int_0^1 \gamma_\varepsilon |u_\varepsilon'|^2.\tag{1.27}$$

Note that in this case $\Omega = (0,1)$, $\chi_\varepsilon(\widehat{x}) = \mathbb{1}_\Omega - (1-\gamma)\mathbb{1}_{\omega_\varepsilon(\widehat{x})}$, with $\omega_\varepsilon(\widehat{x}) = (0,\varepsilon)$, and $\gamma_\varepsilon := \chi_\varepsilon$. It means that $\gamma_\varepsilon(x) = \gamma$ for $0 < x \le \varepsilon$ and $\gamma_\varepsilon(x) = 1$ for $\varepsilon < x < 1$, where $\gamma \in \mathbb{R}^+$, i.e., $0 < \gamma < \infty$, is the contrast on the material property. In addition, u_ε is the solution of the following boundary value problem

$$(\gamma_\varepsilon(x)u_\varepsilon'(x))' = 0, \qquad 0 < x < 1,\tag{1.28}$$

endowed with boundary conditions of the form

$$u_\varepsilon(0) = 0 \qquad \text{and} \qquad u_\varepsilon'(1) = 1,\tag{1.29}$$

and transmission conditions arising naturally from the variational formulation of problem (1.28), that is

$$u_\varepsilon(\varepsilon^+) = u_\varepsilon(\varepsilon^-) \qquad \text{and} \qquad u_\varepsilon'(\varepsilon^+) = \gamma u_\varepsilon'(\varepsilon^-).\tag{1.30}$$

The above boundary value problem admits an explicit solution of the form

$$\begin{cases} u_\varepsilon(x) = \frac{x}{\gamma}, & 0 < x \le \varepsilon, \\ u_\varepsilon(x) = x + \varepsilon\frac{1-\gamma}{\gamma}, & \varepsilon < x < 1. \end{cases} \tag{1.31}$$

Since $|B_\varepsilon| \to 0$ with $\varepsilon \to 0$, there is

$$\psi(\chi) := \int_0^1 |u'|^2, \tag{1.32}$$

where u is the solution to the above boundary value problem for $\varepsilon = 0$, that is

$$u(x) = x. \tag{1.33}$$

We are looking for an asymptotic expansion of the form (1.2), namely

$$\psi(\chi_\varepsilon(\widehat{x})) = \int_0^\varepsilon \gamma|u_\varepsilon'|^2 + \int_\varepsilon^1 |u_\varepsilon'|^2 = 1 + \varepsilon\frac{1-\gamma}{\gamma} = \psi(\chi) + \varepsilon\frac{1-\gamma}{\gamma}. \tag{1.34}$$

Therefore, the associated topological derivative is given by

$$\mathcal{T}(\widehat{x}) = \frac{1-\gamma}{\gamma} \quad \forall\, \widehat{x} \in (0,1), \tag{1.35}$$

with $f(\varepsilon) = \varepsilon$. Note that in the limit case $\gamma \to \infty$, $\mathcal{T}(\widehat{x}) = -1$ $\forall \widehat{x} \in (0,1)$. On the other hand, for $\gamma \to 0$ the topological derivative is not defined. This is in fact an intrinsic property of one-dimensional problems, which in general do not admit singular domain perturbations [18].

1.2 Evaluation of the Topological Derivative

Before concluding this chapter, let us present a last example concerning the simplest case of topological perturbation with PDEs constraints. It is given by a perturbation on the right-hand side of a boundary value problem, which can be seen as a simple variant of the case associated with singularly perturbed geometrical domains. We start by introducing the adjoint sensitivity method. Then we state an auxiliary result which will be used here, in this section, and later in the book. Finally, we present a simple example in the context of optimal control problem.

1.2.1 Adjoint Sensitivity Method

Let us introduce the *adjoint sensitivity method* through the *augmented Lagrangian formalism*. We consider the following minimization problem:

$$\underset{\Omega \in X}{\text{Minimize}} \; \mathscr{J}(u), \tag{1.36}$$

where X represents the set of admissible geometrical domains and $\mathscr{J} : \mathscr{U} \mapsto \mathbb{R}$ is the shape functional to be minimized with respect to the design variable domain $\Omega \subset \mathbb{R}^d$, $d = 2, 3$. In addition, function u is the solution of the abstract variational problem of the form

$$u \in \mathscr{U} \; : \; a(u, \eta) = \ell(\eta) \quad \forall \eta \in \mathscr{V}, \tag{1.37}$$

where $\mathscr{U} \in U$ is the set of admissible functions and $\mathscr{V} \in V$ is the space of admissible variations, with U and V used to denote linear Hilbert subspaces, respectively. Finally, $a : U \times V \mapsto \mathbb{R}$ is a bilinear form and $\ell : V \mapsto \mathbb{R}$ is a linear functional. From these elements, we can introduce the associated augmented Lagrangian, which consists in imposing the constraint of the minimization problem (1.36), given by the state equation (1.37), through Lagrangian multiplier, namely

$$\mathscr{L}(u, v) = \mathscr{J}(u) + a(u, v) - \ell(v) \quad \forall (u, v) \in \mathscr{U} \times \mathscr{V}. \tag{1.38}$$

Let us evaluate the *Fréchet derivative* of the Lagrangian function $\mathscr{L}(u, v)$ with respect to $v \in \mathscr{V}$ in the direction $\eta \in \mathscr{V}$, thus

$$\langle D_v \mathscr{L}(u, v), \eta \rangle = a(u, \eta) - \ell(\eta). \tag{1.39}$$

After applying the first order optimality condition in the above result we obtain

$$u \in \mathscr{U} \; : \; a(u, \eta) = \ell(\eta) \quad \forall \eta \in \mathscr{V}, \tag{1.40}$$

which is actually the state equation (1.37). On the other hand, the *Fréchet derivative* of the Lagrangian function $\mathscr{L}(u, v)$ with respect to $u \in \mathscr{U}$ in the direction $\varphi \in \mathscr{V}$ can be written as

$$\langle D_u \mathscr{L}(u, v), \varphi \rangle = \langle D_u \mathscr{J}(u), \varphi \rangle + a(\varphi, v). \tag{1.41}$$

Let us apply again the first order optimality condition, leading to the associated *adjoint equation*, namely

$$v \in \mathscr{V} \; : \; a(\varphi, v) = -\langle D_u \mathscr{J}(u), \varphi \rangle \quad \forall \varphi \in \mathscr{V}. \tag{1.42}$$

Note that, from (1.42), the adjoint state v always lives in the space $\mathscr{V}$ and appears on the second argument of the bilinear form. Finally, from the above discussion, the adjoint variable v can also be interpreted as the Lagrangian multiplier used to impose the state equation (1.37) as a constraint in the optimization problem (1.36).

1.2.2 Auxiliary Result

Now, let us state an auxiliary result which will be used in the next section in particular and in the whole book in general.

Lemma 1.1 *Let Ω be an open and bounded domain in $\mathbb{R}^2$ and let B_ε be a ball of radius ε, such that $\overline{B_\varepsilon} \subset \Omega$. Then, for a function $\varphi \in H^1(\Omega)$, the following estimate holds true*

$$\|\varphi\|_{L^2(B_\varepsilon)} \leq C\varepsilon^\delta \|\varphi\|_{H^1(\Omega)}, \tag{1.43}$$

with $0 < \delta < 1$ and the constant C independent of the small parameter ε.

Proof From the *Hölder inequality*, we have

$$\|\varphi\|_{L^2(B_\varepsilon)} \leq \left[\left(\int_{B_\varepsilon} (|\varphi|^2)^p \right)^{\frac{1}{p}} \left(\int_{B_\varepsilon} 1^q \right)^{\frac{1}{q}} \right]^{\frac{1}{2}}$$

$$= \pi^{1/2q} \varepsilon^{1/q} \left(\int_{B_\varepsilon} |\varphi|^{2p} \right)^{\frac{1}{2p}}$$

$$= \pi^{1/2q} \varepsilon^{1/q} \|\varphi\|_{L^{2p}(B_\varepsilon)}$$

$$\leq C\varepsilon^{1/q} \|\varphi\|_{L^{2p}(\Omega)}, \tag{1.44}$$

for all $p, q \in (1, +\infty)$ satisfying $1/p + 1/q = 1$. By choosing $q > 1$ and p accordingly, the *Sobolev embedding theorem* [33, Ch. IV, §8, Sec. 1.2, p. 139] implies $H^1(\Omega) \subset L^{2p}(\Omega)$ with a continuous embedding. Therefore, we have

$$\|\varphi\|_{L^2(B_\varepsilon)} \leq C\varepsilon^{1/q} \|\varphi\|_{H^1(\Omega)}, \tag{1.45}$$

which leads to the result by setting $\delta = 1/q$. $\square$

1.2.3 A Simple Example

Let us consider the tracking type *shape functional*, which is useful in many practical applications including optimal control problem and imaging processing, namely

$$\psi(\chi) := \mathscr{J}(u) = \frac{1}{2}\int_{\Omega}|u - z_d|^2, \qquad (1.46)$$

where $\Omega \subset \mathbb{R}^2$ and z_d is the target function, assumed to be smooth. The scalar field u is the solution of the following variation problem:

$$u \in H_0^1(\Omega) : \int_{\Omega}\nabla u \cdot \nabla \eta = \int_{\Omega}b\eta \quad \forall \eta \in H_0^1(\Omega), \qquad (1.47)$$

where the source-term b is assumed to be locally Lipschitz continuous (see Example 1.2). According to Sect. 1.2.1, the associated augmented Lagrangian functional is given by

$$\mathscr{L}(u, v) = \frac{1}{2}\int_{\Omega}|u - z_d|^2 + \int_{\Omega}\nabla u \cdot \nabla v - \int_{\Omega}bv, \qquad (1.48)$$

and the *adjoint equation* reads

$$v \in H_0^1(\Omega) : \int_{\Omega}\nabla \eta \cdot \nabla v = -\langle D_u \mathscr{J}(u), \eta\rangle$$

$$= -\int_{\Omega}(u - z_d)\eta \quad \forall \eta \in H_0^1(\Omega). \qquad (1.49)$$

Note that by symmetry of the above bilinear form, in this particular case the left-hand sides of (1.47) and (1.49) are precisely the same but for the appropriate test functions only. The only difference is their right-hand sides. This property simplifies enormously the numerics.

Now, we introduce a topological perturbation on the source term of the form $b_\varepsilon = \chi_\varepsilon b$, with $\chi_\varepsilon(\widehat{x}) = \mathbb{1}_\Omega - (1 - \gamma)\mathbb{1}_{B_\varepsilon(\widehat{x})}$. Therefore, the perturbed source term b_ε can be written as

$$b_\varepsilon(x) := \begin{cases} b(x) \text{ if } x \in \Omega \setminus B_\varepsilon(\widehat{x}), \\ \gamma b(x) \text{ if } x \in B_\varepsilon(\widehat{x}), \end{cases} \qquad (1.50)$$

with $\gamma \in \mathbb{R}$ used to denote the contrast in the source term. From these elements, the shape functional associated with the perturbed problem is defined as

$$\psi(\chi_\varepsilon) := \mathscr{J}_\varepsilon(u_\varepsilon) = \frac{1}{2}\int_{\Omega}|u_\varepsilon - z_d|^2. \qquad (1.51)$$

The scalar function u_ε is the solution of the following variation problem

$$u_\varepsilon \in H_0^1(\Omega) : \int_\Omega \nabla u_\varepsilon \cdot \nabla \eta = \int_\Omega b_\varepsilon \eta \quad \forall \eta \in H_0^1(\Omega). \tag{1.52}$$

From the definition of the source term given by (1.50), we have $b_\varepsilon = b$ in $\Omega \setminus \overline{B_\varepsilon}$ and $b_\varepsilon = \gamma b$ in B_ε. Therefore, the state equation (1.52) can be rewritten as

$$u_\varepsilon \in H_0^1(\Omega) : \int_\Omega \nabla u_\varepsilon \cdot \nabla \eta = \int_{\Omega \setminus \overline{B_\varepsilon}} b\eta + \gamma \int_{B_\varepsilon} b\eta \pm \int_{B_\varepsilon} b\eta$$

$$= \int_\Omega b\eta - (1-\gamma) \int_{B_\varepsilon} b\eta \quad \forall \eta \in H_0^1(\Omega). \tag{1.53}$$

Now, let us subtract (1.47) from (1.53) to obtain

$$\int_\Omega \nabla(u_\varepsilon - u) \cdot \nabla \eta = -(1-\gamma) \int_{B_\varepsilon} b\eta \quad \forall \eta \in H_0^1(\Omega). \tag{1.54}$$

Thus, the existence of the topological derivative of the problem we are dealing with is ensured by the following result:

Lemma 1.2 *Let u and u_ε be the solutions of* (1.47) *and* (1.52), *respectively. Then the following estimate holds true*

$$\|u_\varepsilon - u\|_{H^1(\Omega)} \le C\varepsilon^{1+\delta}, \tag{1.55}$$

with constant C independent of the small parameter ε and $0 < \delta < 1$.

Proof By taking $\eta = u_\varepsilon - u$ as test function in (1.54), we obtain the following equality:

$$\int_\Omega \|\nabla(u_\varepsilon - u)\|^2 = -(1-\gamma) \int_{B_\varepsilon} b(u_\varepsilon - u). \tag{1.56}$$

From the *Cauchy-Schwarz inequality*, we have

$$\int_\Omega \|\nabla(u_\varepsilon - u)\|^2 \le C_1 \|b\|_{L^2(B_\varepsilon)} \|u_\varepsilon - u\|_{L^2(B_\varepsilon)}$$

$$\le C_2 \varepsilon^{1+\delta} \|u_\varepsilon - u\|_{H^1(\Omega)},$$

where we have used Lemma 1.1 and the continuity of function b at the point $\widehat{x} \in \Omega$. Finally, from the *coercivity* of the bilinear form on the left-hand side of the above inequality, namely

$$c\|u_\varepsilon - u\|_{H^1(\Omega)}^2 \le \int_\Omega \|\nabla(u_\varepsilon - u)\|^2, \tag{1.57}$$

we conclude that

$$c\|u_\varepsilon - u\|_{H^1(\Omega)} \le C_2\varepsilon^{1+\delta}, \tag{1.58}$$

which leads to the result with $C = C_2/c$ and $0 < \delta < 1$. $\square$

The variation of the shape functional can be obtained by subtracting (1.46) from (1.51), that is

$$\mathcal{J}_\varepsilon(u_\varepsilon) - \mathcal{J}(u) = \int_\Omega (u - z_d)(u_\varepsilon - u) + \mathcal{E}_1(\varepsilon), \tag{1.59}$$

with the remainder $\mathcal{E}_1(\varepsilon)$ bounded as

$$\mathcal{E}_1(\varepsilon) = \frac{1}{2}\int_\Omega |u_\varepsilon - u|^2,$$

$$|\mathcal{E}_1(\varepsilon)| \le C\|u_\varepsilon - u\|_{L^2(\Omega)}^2$$

$$\le C\|u_\varepsilon - u\|_{H^1(\Omega)}^2 = o(\varepsilon^2), \tag{1.60}$$

where we have used Lemma 1.2. Now, let us set $\eta = u_\varepsilon - u$ as test function in the adjoint Eq. (1.49) and $\eta = v$ as test function in (1.54) to obtain

$$\int_\Omega \nabla v \cdot \nabla(u_\varepsilon - u) = -\int_\Omega (u - z_d)(u_\varepsilon - u), \tag{1.61}$$

$$\int_\Omega \nabla v \cdot \nabla(u_\varepsilon - u) = -(1 - \gamma)\int_{B_\varepsilon} bv. \tag{1.62}$$

From the above results we conclude that

$$\int_\Omega (u - z_d)(u_\varepsilon - u) = (1 - \gamma)\int_{B_\varepsilon} bv. \tag{1.63}$$

Therefore, the variation (1.59) can be rewritten as an integral concentrated in the ball B_ε, namely

$$\mathcal{J}_\varepsilon(u_\varepsilon) - \mathcal{J}(u) = (1 - \gamma)\int_{B_\varepsilon} bv + \mathcal{E}_1(\varepsilon)$$

$$= (1 - \gamma)\pi\varepsilon^2 b(\widehat{x})v(\widehat{x}) + \mathcal{E}_1(\varepsilon) + \mathcal{E}_2(\varepsilon). \tag{1.64}$$

The remainder $\mathcal{E}_2(\varepsilon)$ can be bounded as follows:

$$\mathcal{E}_2(\varepsilon) = (1 - \gamma)\int_{B_\varepsilon} (bv - b(\widehat{x})v(\widehat{x})),$$

$$|\mathscr{E}_2(\varepsilon)| \leq C_1 \int_{B_\varepsilon} \|x - \widehat{x}\| \leq C_2 \varepsilon^3 = o(\varepsilon^2), \qquad (1.65)$$

where we have used the interior elliptic regularity of function u. See Example 1.2. Finally, the topological asymptotic expansion of the shape functional is given by

$$\mathscr{J}_\varepsilon(u_\varepsilon) = \mathscr{J}(u) + \pi \varepsilon^2 (1 - \gamma) b(\widehat{x}) v(\widehat{x}) + o(\varepsilon^2). \qquad (1.66)$$

From the above expansion, we can identify function $f(\varepsilon) = \pi \varepsilon^2$ and the final formula for the topological derivative, namely

$$\mathscr{T}(\widehat{x}) = (1 - \gamma) b(\widehat{x}) v(\widehat{x}) \quad \forall \widehat{x} \in \Omega, \qquad (1.67)$$

where u and v are the solutions of the direct (1.47) and adjoint (1.49) problems, respectively, both defined in the original (unperturbed) domain Ω.

Remark 1.3 This kind of topological perturbation, that is, on the right-hand side of the governing boundary value problem, can be treated by using simple arguments from the analysis. Actually, we have just used the fact that the boundary value problems are well posed. Therefore, in this context, it is possible to consider certain classes of nonlinear problems. However, this book is dedicated to the case of topological perturbations on the main part of the differential operator, such as the ones produced by the nucleation of holes. The mathematical analysis of this class of topological perturbations is much more involved, which is deeply discussed in Chap. 4 and also in [75], for instance.

1.3   Organization of the Book

In this chapter the topological derivative concept has been introduced, together with some selected examples. Note that the small parameter governing the asymptotic analysis represents the size of the topological domain perturbation, allowing for the nucleation of small inclusions or voids in a numerical procedure of optimization regarding shape/topology changes on the material properties distribution or in the geometrical domain itself, respectively. Therefore, this new concept in shape optimization has applications in many different fields such as topology optimization, inverse problems, imaging processing, multi-scale material design, and mechanical modeling including damage and fracture evolution phenomena. The central idea of this work is to introduce the topological derivative method from both theoretical and practical point of views, so that it is oriented to the readers interested in the mathematical aspects of the topological asymptotic analysis as well as in the applications of the topological derivative method in computational mechanics. In particular, this book is presented as follows:

- The topological asymptotic analysis of the energy shape functional associated with the Poisson's equation, with respect to singular domain perturbations, is formally developed through Chap. 2. In particular, we consider singular perturbations produced by the nucleation of small circular holes endowed with homogeneous Neumann, Dirichlet, or Robin boundary conditions.
- Chapter 3 deals with the topological derivative of the so-called compliance shape functional associated with a modified Helmholtz problem, with respect to the nucleation of a small circular inclusion with different material property from the background. By taking into account the boundary value problem we are dealing with, three different cases are considered: (1) perturbation on its right-hand side, (2) perturbation on the lower order term, and (3) perturbation on the higher order term. The existence of the associated topological derivatives is ensured by using simple arguments from the analysis. Then, we derive their explicit forms which are useful for numerical methods in shape/topology optimization. Finally, a priori estimates for the remainders left in the topological asymptotic expansions are rigorously obtained, which are used to justify the obtained results.
- The domain decomposition technique combined with the Steklov–Poincaré pseudo-differential boundary operator is presented in Chap. 4. The main idea is introduced in the context of coupled elliptic boundary value problems. Then, the same framework is used for deriving the topological asymptotic expansion of a tracking-type shape functional with respect to singular domain perturbations produced by the nucleation of small circular holes endowed with homogeneous Neumann boundary condition. The resulting asymptotic method allows for obtaining sharp a priori estimates for the remainders, so that it can be seen as a rigorous mathematical justification for the derivations presented in the former chapters.
- In Chap. 5 a topology optimization algorithm based on the topological derivative concept combined with a level-set domain representation method is presented. The model problem is governed by the elasticity system into two spatial dimensions. The topological asymptotic expansion of a tracking-type shape functional, associated with the nucleation of a small circular inclusion endowed with different material property from the background, is rigorously derived. Finally, the obtained theoretical result is used as a steepest-descent direction in the optimization process, which is applied in the context of compliance structural topology optimization and topology design of compliant mechanisms.
- Some useful basic results of tensor calculus are included in Appendix A for the readers' convenience. In particular, inner, vector, and tensor products are defined. In addition, gradient, divergence, and curl formulae, together with some integral theorems, are presented. Finally, some useful decompositions in curvilinear, polar, and spherical coordinate systems are provided.

1.4 Exercises

1. Repeat Example 1.2 by considering $g : \mathbb{R}^3 \mapsto \mathbb{R}$.
2. Let us consider one more term in the topological asymptotic expansion of the form

$$\psi(\chi_\varepsilon(\widehat{x})) = \psi(\chi) + f(\varepsilon)\mathcal{T}(\widehat{x}) + f_2(\varepsilon)\mathcal{T}^2(\widehat{x}) + o(f_2(\varepsilon)),$$

where $f_2(\varepsilon)$ is such that

$$\lim_{\varepsilon \to 0} \frac{f_2(\varepsilon)}{f(\varepsilon)} = 0.$$

Then, quantities $\mathcal{T}(\widehat{x})$ and $\mathcal{T}^2(\widehat{x})$ represent the first and *second order topological derivatives* of ψ, respectively. Assume that function $g(x)$ in Example 1.3 is of class $C^2(\Omega)$, with its second order gradient Lipschitz continuous in B_ε, namely $\exists\, C \geq 0 : \|\nabla\nabla g(x) - \nabla\nabla g(\widehat{x})\| \leq C\|x - \widehat{x}\|\ \forall x \in B_\varepsilon$, where $B_\varepsilon(\widehat{x})$ is used to denote a ball of radius ε and center at $\widehat{x} \in \Omega \subset \mathbb{R}^2$. Show that the topological asymptotic expansion of the functional $\psi(\chi_\varepsilon(\widehat{x}))$ is given by

$$\psi(\chi_\varepsilon(\widehat{x})) = \psi(\chi) - (1-\gamma)\pi\varepsilon^2 g(\widehat{x}) - \frac{1-\gamma}{8}\pi\varepsilon^4 \Delta g(\widehat{x}) + o(\varepsilon^4).$$

3. From the definition for the topological asymptotic expansion given by (1.2), show the results presented in Example 1.4.
4. Repeat Example 1.5 by considering the following conditions:

$$-u''(x) = 1 \quad \text{for} \quad 0 < x < 1 \quad \text{and} \quad u(0) = u'(1) = 0.$$

5. By taking into account the example presented in Sect. 1.2.3:

   (a) Replace the shape functional (1.46) by the total potential energy associated with the variational problem (1.47), namely

$$\psi(\chi) := \mathscr{J}(u) = \frac{1}{2}\int_\Omega \|\nabla u\|^2 - \int_\Omega bu.$$

   Then compute its topological derivative by repeating the same derivations as presented in Sect. 1.2.3.

   (b) Take the total potential energy defined in a disk B_1 of unit radius and center at the origin. By setting $b = 1$ as source term, consider as topological perturbation the particular case given by (1.50), namely $b_\varepsilon(x) = 1$ if $x \in B_1 \setminus \overline{B_\varepsilon}$ and $b_\varepsilon(x) = \gamma$ if $x \in B_\varepsilon$, where B_ε is a disk of radius ε and center at the origin. From these elements, develop the topologically perturbed counter part of the total potential energy given by $\psi(\chi_\varepsilon)$ in power of ε around the origin. Finally, compare the obtained result with the one previously derived.

Chapter 2
Singular Domain Perturbation

In this chapter, the topological derivative of the total potential energy associated with the Poisson's problem is formally evaluated by considering homogeneous Neumann and Dirichlet conditions on the boundary of the hole ∂B_ε. The mathematical justification for the derived results can be found in Chap. 4 and also in the book by Novotny and Sokolowski [75], for instance. In this case, the geometrical domain is topologically perturbed by the nucleation of a small circular hole, as shown in Fig. 2.1. Since $\Omega \subset \mathbb{R}^2$ is the original (unperturbed) domain, then $\Omega_\varepsilon(\widehat{x}) = \Omega \setminus \overline{B_\varepsilon(\widehat{x})}$ is the topologically perturbed domain, where $B_\varepsilon(\widehat{x})$, with $\overline{B_\varepsilon} \subset \Omega$, is used to denote a ball of radius ε, $0 < \varepsilon < \mathrm{dist}(\widehat{x}, \partial\Omega)$, and center at $\widehat{x} \in \Omega$.

2.1 Problem Formulation

The *shape functional* associated with the unperturbed domain Ω is defined as

$$\psi(\chi) := \mathscr{J}_\Omega(u) = \frac{1}{2}\int_\Omega \|\nabla u\|^2 - \int_\Omega bu, \qquad (2.1)$$

where the scalar function $u : \Omega \mapsto \mathbb{R}$ is the solution of the following variational problem

$$u \in H_0^1(\Omega) : \int_\Omega \nabla u \cdot \nabla \eta = \int_\Omega b\eta \quad \forall \eta \in H_0^1(\Omega). \qquad (2.2)$$

In the above equation, b is a source term considered constant everywhere. The strong form associated with the variational problem (2.2) is given by the following boundary value problem governed by the Poisson's equation: Find u, such that

A. A. Novotny, J. Sokołowski, *An Introduction to the Topological Derivative Method*, SpringerBriefs in Mathematics, https://doi.org/10.1007/978-3-030-36915-6_2

17

Fig. 2.1 Topologically
perturbed domain by the
nucleation of a small circular
hole

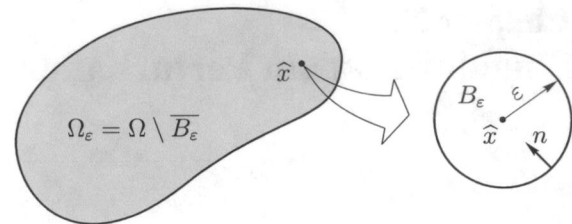

$$\begin{cases} -\Delta u = b \text{ in } \Omega, \\ \quad u = 0 \text{ on } \partial\Omega. \end{cases} \tag{2.3}$$

Let us consider the same problem but now defined in the topologically perturbed domain Ω_ε, so that the associated total potential energy is given by

$$\psi(\chi_\varepsilon) := \mathscr{J}_{\Omega_\varepsilon}(u_\varepsilon) = \frac{1}{2}\int_{\Omega_\varepsilon} \|\nabla u_\varepsilon\|^2 - \int_{\Omega_\varepsilon} b u_\varepsilon, \tag{2.4}$$

where the scalar function $u_\varepsilon : \Omega_\varepsilon \mapsto \mathbb{R}$ is the solution of the following variational problem

$$u_\varepsilon \in \mathscr{V}_\varepsilon : \int_{\Omega_\varepsilon} \nabla u_\varepsilon \cdot \nabla \eta = \int_{\Omega_\varepsilon} b\eta \quad \forall \eta \in \mathscr{V}_\varepsilon. \tag{2.5}$$

The space $\mathscr{V}_\varepsilon$ is defined according to the boundary condition on ∂B_ε, namely

$$\mathscr{V}_\varepsilon := \{\varphi \in H^1(\Omega_\varepsilon) : \varphi_{|\partial\Omega} = 0, \ \beta\varphi_{|\partial B_\varepsilon} = 0\}, \tag{2.6}$$

with $\beta \in \{0, 1\}$. This notation has to be understood as follows:

- For $\beta = 1$, we have homogeneous Dirichlet condition on the boundary of the ball ∂B_ε, since in this case $u_\varepsilon = 0$ and $\eta = 0$ on ∂B_ε.
- For $\beta = 0$, u_ε and η are free on the boundary of the hole ∂B_ε, leading to homogeneous Neumann boundary condition on ∂B_ε.

Finally, the *strong form* associated with the variational problem (2.5) can be stated as: Find u_ε, such that

$$\begin{cases} -\Delta u_\varepsilon = b \text{ in } \Omega_\varepsilon, \\ \qquad u_\varepsilon = 0 \text{ on } \partial\Omega, \\ \beta u_\varepsilon + (1-\beta)\partial_n u_\varepsilon = 0 \text{ on } \partial B_\varepsilon. \end{cases} \tag{2.7}$$

2.2   Variation of the Energy Shape Functional

Through a simple calculation, it is possible to show that the variation of the total potential energy leads to integrals concentrated in the ball B_ε as well as on its boundary ∂B_ε. In fact, by taking $\eta = u$ in (2.2) and $\eta = u_\varepsilon$ in (2.5), we obtain respectively the following equalities:

$$\frac{1}{2}\int_\Omega \|\nabla u\|^2 = \frac{1}{2}\int_\Omega bu \quad \text{and} \quad \frac{1}{2}\int_{\Omega_\varepsilon} \|\nabla u_\varepsilon\|^2 = \frac{1}{2}\int_{\Omega_\varepsilon} bu_\varepsilon, \tag{2.8}$$

which allow for rewriting $\mathscr{J}_\Omega(u)$ and $\mathscr{J}_{\Omega_\varepsilon}(u_\varepsilon)$ as

$$\mathscr{J}_\Omega(u) = -\frac{1}{2}\int_{\Omega_\varepsilon} bu - \frac{1}{2}\int_{B_\varepsilon} bu \quad \text{and} \quad \mathscr{J}_{\Omega_\varepsilon}(u_\varepsilon) = -\frac{1}{2}\int_{\Omega_\varepsilon} bu_\varepsilon, \tag{2.9}$$

where $\Omega_\varepsilon = \Omega \setminus \overline{B_\varepsilon}$. Therefore, the variation of the total potential reads

$$\mathscr{J}_{\Omega_\varepsilon}(u_\varepsilon) - \mathscr{J}_\Omega(u) = \frac{1}{2}\int_{B_\varepsilon} bu - \frac{1}{2}\int_{\Omega_\varepsilon} b(u_\varepsilon - u). \tag{2.10}$$

On the other hand, after rewriting (2.2) as follows:

$$\int_{\Omega_\varepsilon} \nabla u \cdot \nabla \eta + \int_{B_\varepsilon} \nabla u \cdot \nabla \eta = \int_{\Omega_\varepsilon} b\eta + \int_{B_\varepsilon} b\eta, \tag{2.11}$$

the divergence theorem yields (see Appendix A, identity (A.42))

$$\int_{\Omega_\varepsilon} \nabla u \cdot \nabla \eta - \int_{B_\varepsilon} (\Delta u + b)\eta - \int_{\partial B_\varepsilon} \partial_n u \eta = \int_{\Omega_\varepsilon} b\eta, \tag{2.12}$$

where n is the unit normal vector field on ∂B_ε pointing toward to the center of the hole, as shown in Fig. 2.1. Since u is the solution of (2.3), namely $\Delta u + b = 0$, we have

$$\int_{\Omega_\varepsilon} \nabla u \cdot \nabla \eta - \int_{\partial B_\varepsilon} \partial_n u \eta = \int_{\Omega_\varepsilon} b\eta, \tag{2.13}$$

which allows us to set $\eta = u_\varepsilon$ as test function in the above equation, leading to the following equality:

$$\int_{\Omega_\varepsilon} \nabla u \cdot \nabla u_\varepsilon = \int_{\Omega_\varepsilon} bu_\varepsilon + \int_{\partial B_\varepsilon} u_\varepsilon \partial_n u. \tag{2.14}$$

By applying again the divergence theorem, we obtain

$$\int_{\Omega_\varepsilon} \nabla u_\varepsilon \cdot \nabla u = -\int_{\Omega_\varepsilon} \Delta u_\varepsilon u + \int_{\partial\Omega} u \partial_n u_\varepsilon + \int_{\partial B_\varepsilon} u \partial_n u_\varepsilon$$

$$= \int_{\Omega_\varepsilon} bu + \int_{\partial B_\varepsilon} u \partial_n u_\varepsilon, \tag{2.15}$$

where we have used the fact that u_ε is the solution of (2.7), that is, $-\Delta u_\varepsilon = b$, and that $u = 0$ on $\partial\Omega$. After comparing these last two results, we observe that

$$\int_{\Omega_\varepsilon} b(u_\varepsilon - u) = \int_{\partial B_\varepsilon} (u \partial_n u_\varepsilon - u_\varepsilon \partial_n u). \tag{2.16}$$

Thus, the variation of the total potential energy can be, in fact, written in the form of integrals concentrated in B_ε and on its boundary ∂B_ε, namely

$$\mathcal{J}_{\Omega_\varepsilon}(u_\varepsilon) - \mathcal{J}_\Omega(u) = -\frac{1}{2} \int_{\partial B_\varepsilon} (u \partial_n u_\varepsilon - u_\varepsilon \partial_n u) + \frac{1}{2} \int_{B_\varepsilon} bu. \tag{2.17}$$

From the interior elliptic regularity of the solution u, the last term of (2.17) can be trivially expanded in power of ε as follows (see Chap. 1, Example 1.2)

$$\int_{B_\varepsilon} bu = \pi \varepsilon^2 bu(\widehat{x}) + o(\varepsilon^2), \tag{2.18}$$

which leads to

$$\mathcal{J}_{\Omega_\varepsilon}(u_\varepsilon) - \mathcal{J}_\Omega(u) = -\frac{1}{2} \int_{\partial B_\varepsilon} (u \partial_n u_\varepsilon - u_\varepsilon \partial_n u) + \frac{1}{2}\pi \varepsilon^2 bu(\widehat{x}) + o(\varepsilon^2). \tag{2.19}$$

2.3 Topological Derivative Evaluation

The variation of the total potential energy is given by integrals concentrated in B_ε and on its boundary ∂B_ε. Thus, in order to obtain the associated topological asymptotic expansion in the form (1.2), we have to know the asymptotic behavior of the solution u_ε with respect to ε in the neighborhood of the hole B_ε. In particular, once this behavior is known explicitly, it is possible to identify the function $f(\varepsilon)$ and then evaluate the limit passage $\varepsilon \to 0$ in (1.4) in order to obtain the final formula for the topological derivative $\mathcal{T}$ of the shape functional ψ. However, in general it is not an easy task. Actually, we have to perform an asymptotic analysis of u_ε with respect to the small parameter ε governing the singular domain perturbation.

In this section, we present a formal asymptotic expansion of the solution u_ε regarding both homogenous Neumann and Dirichlet boundary conditions on ∂B_ε. A rigorous mathematical justification for the results to be derived can be found in the book by Kozlov et al. [58], for instance. The reader interested in the general theory of asymptotic analysis of solutions in singularly perturbed geometrical domains may refer to the book by Mazja et al. [68].

Now, we consider the two cases under analysis separately, namely Neumann and Dirichlet boundary conditions on the hole. We will see that the asymptotic behavior of the solution and, consequently, of the shape functional depends strongly on these boundary conditions, leading to completely different asymptotic expansions.

2.3.1 Neumann Boundary Condition on the Hole

By setting $\beta = 0$ in (2.7), there is homogeneous Neumann boundary condition on ∂B_ε, that is, $\partial_n u_\varepsilon |_{\partial B_\varepsilon} = 0$. In this case, the variation of the total potential energy given by (2.19) results in

$$\mathscr{J}_{\Omega_\varepsilon}(u_\varepsilon) - \mathscr{J}_\Omega(u) = \frac{1}{2} \int_{\partial B_\varepsilon} u_\varepsilon \partial_n u + \frac{1}{2} \pi \varepsilon^2 b u(\widehat{x}) + o(\varepsilon^2). \tag{2.20}$$

Now, let us propose an *ansatz* for the expansion of u_ε in the following form [58, 68]:

$$u_\varepsilon(x) = u(x) + \varepsilon w(\varepsilon^{-1}x) + \widetilde{u}_\varepsilon(x)$$

$$= u(\widehat{x}) + \nabla u(\widehat{x}) \cdot (x - \widehat{x}) + \frac{1}{2} \nabla \nabla u(y)(x - \widehat{x}) \cdot (x \quad \widehat{x})$$

$$+ \varepsilon w(\varepsilon^{-1}x) + \widetilde{u}_\varepsilon(x), \tag{2.21}$$

where y is an intermediate point between x and $\widehat{x}$. Since on the boundary of the hole ∂B_ε there is $\partial_n u_\varepsilon |_{\partial B_\varepsilon} = 0$, the normal derivative of the above expansion, evaluated on ∂B_ε, leads to

$$\nabla u(\widehat{x}) \cdot n - \varepsilon \nabla \nabla u(y)n \cdot n + \varepsilon \partial_n w(\varepsilon^{-1}x) + \partial_n \widetilde{u}_\varepsilon(x) = 0. \tag{2.22}$$

Let us choose w such that

$$\varepsilon \partial_n w(\varepsilon^{-1}x) = -\nabla u(\widehat{x}) \cdot n \quad \text{on} \quad \partial B_\varepsilon. \tag{2.23}$$

In the fast variable $\xi = \varepsilon^{-1}x$, which implies $\nabla_\xi w(\xi) = \varepsilon \nabla w(\varepsilon^{-1}x)$, the following *exterior problem* is formally defined for $\varepsilon \to 0$: Find w, such that

$$
\begin{cases}
\Delta_\xi w = 0 & \text{in } \mathbb{R}^2 \setminus \overline{B_1}, \\
w \to 0 & \text{at } \infty, \\
\nabla_\xi w \cdot n = -\nabla u(\widehat{x}) \cdot n & \text{on } \partial B_1.
\end{cases}
\tag{2.24}
$$

The above boundary value problem admits an *explicit solution*, namely

$$
w(\varepsilon^{-1}x) = \frac{\varepsilon}{\|x - \widehat{x}\|^2} \nabla u(\widehat{x}) \cdot (x - \widehat{x}).
\tag{2.25}
$$

The general solution for the Laplace equation into two spatial dimensions written in terms of Fourier series can be found in Appendix A, Eq. (A.74). Now, we have to construct $\widetilde{u}_\varepsilon$ is such a way that it compensates for the discrepancies left by the higher order terms in ε as well as by the boundary layer w on the exterior boundary $\partial \Omega$. That is, the *remainder* $\widetilde{u}_\varepsilon$ has to be the solution of the following boundary value problem: Find $\widetilde{u}_\varepsilon$, such that

$$
\begin{cases}
\Delta \widetilde{u}_\varepsilon = 0 & \text{in } \Omega_\varepsilon, \\
\widetilde{u}_\varepsilon = -\varepsilon w & \text{on } \partial \Omega, \\
\partial_n \widetilde{u}_\varepsilon = \varepsilon \nabla \nabla u(y) n \cdot n & \text{on } \partial B_\varepsilon.
\end{cases}
\tag{2.26}
$$

Under suitable regularity conditions, the estimate $\widetilde{u}_\varepsilon \approx O(\varepsilon^2)$ holds true in an appropriate norm [58, 68]. Therefore, the *expansion* for u_ε can be written as

$$
u_\varepsilon(x) = u(x) + \frac{\varepsilon^2}{\|x - \widehat{x}\|^2} \nabla u(\widehat{x}) \cdot (x - \widehat{x}) + O(\varepsilon^2).
\tag{2.27}
$$

Let us introduce the above ansatz into the first term of (2.20) to obtain

$$
\int_{\partial B_\varepsilon} u_\varepsilon \partial_n u = \int_{\partial B_\varepsilon} u \partial_n u + \varepsilon \int_{\partial B_\varepsilon} w(\varepsilon^{-1}x) \partial_n u + o(\varepsilon^2).
\tag{2.28}
$$

After applying the divergence theorem (see Appendix A, identity (A.42)), and from the interior elliptic regularity of the solution u, we have

$$
\int_{\partial B_\varepsilon} u \partial_n u = -\int_{B_\varepsilon} \|\nabla u\|^2 - \int_{B_\varepsilon} u \Delta u = -\int_{B_\varepsilon} \|\nabla u\|^2 + \int_{B_\varepsilon} bu
$$

$$
= -\pi \varepsilon^2 \|\nabla u(\widehat{x})\|^2 + \pi \varepsilon^2 bu(\widehat{x}) + o(\varepsilon^2),
\tag{2.29}
$$

where the normal n points toward the center of the hole. On the other hand, the normal derivative of u evaluated on ∂B_ε can be expanded in Taylor series as follows:

$$\partial_n u(x)|_{\partial B_\varepsilon} = \nabla u(\widehat{x}) \cdot n - \varepsilon \nabla \nabla u(y) n \cdot n$$
$$= \nabla u(\widehat{x}) \cdot n + O(\varepsilon), \tag{2.30}$$

where y is an intermediate point between x and $\widehat{x}$. Moreover, function $w(\varepsilon^{-1}x)$ written in its explicit form through (2.25) can also be evaluated on the boundary of the hole ∂B_ε, which results in

$$w(\varepsilon^{-1}x)|_{\partial B_\varepsilon} = -\nabla u(\widehat{x}) \cdot n, \tag{2.31}$$

since $\|x - \widehat{x}\| = \varepsilon$ and $x - \widehat{x} = -\varepsilon n$ on ∂B_ε. Thus,

$$\varepsilon \int_{\partial B_\varepsilon} w(\varepsilon^{-1}x)\partial_n u = -\varepsilon \int_{\partial B_\varepsilon} (\nabla u(\widehat{x}) \cdot n)(\nabla u(\widehat{x}) \cdot n) + o(\varepsilon^2)$$
$$= -\varepsilon(\nabla u(\widehat{x}) \otimes \nabla u(\widehat{x})) \cdot \int_{\partial B_\varepsilon} n \otimes n + o(\varepsilon^2)$$
$$= -\pi\varepsilon^2(\nabla u(\widehat{x}) \otimes \nabla u(\widehat{x})) \cdot I + o(\varepsilon^2)$$
$$= -\pi\varepsilon^2(\nabla u(\widehat{x}) \cdot \nabla u(\widehat{x})) + o(\varepsilon^2)$$
$$= -\pi\varepsilon^2\|\nabla u(\widehat{x})\|^2 + o(\varepsilon^2). \tag{2.32}$$

From the above results, the first term in (2.20) can be expanded in power of ε as follows:

$$\int_{\partial B_\varepsilon} u_\varepsilon \partial_n u = -2\pi\varepsilon^2\|\nabla u(\widehat{x})\|^2 + \pi\varepsilon^2 bu(\widehat{x}) + o(\varepsilon^2). \tag{2.33}$$

This last result together with expansion (2.20) allows for writing the variation of the total potential energy in the following form:

$$\mathscr{J}_{\Omega_\varepsilon}(u_\varepsilon) - \mathscr{J}_\Omega(u) = -\pi\varepsilon^2\left(\|\nabla u(\widehat{x})\|^2 - bu(\widehat{x})\right) + o(\varepsilon^2). \tag{2.34}$$

Now, in order to identify the leading term in the above expansion, we take

$$f(\varepsilon) = \pi\varepsilon^2, \tag{2.35}$$

leading to the final formula for the *topological derivative* [76, 88]

$$\mathscr{T}(\widehat{x}) = -\|\nabla u(\widehat{x})\|^2 + bu(\widehat{x}) \quad \forall\, \widehat{x} \in \Omega. \tag{2.36}$$

From this procedure, the *topological asymptotic expansion* of the energy shape function can be written as

$$\psi(\chi_\varepsilon(\widehat{x})) = \psi(\chi) - \pi\varepsilon^2(\|\nabla u(\widehat{x})\|^2 - bu(\widehat{x})) + o(\varepsilon^2). \tag{2.37}$$

The complete mathematical justification for the above expansion can be found in Chap. 4, as well as in the book by Novotny and Sokolowski [75, Ch. 10], for instance. In order to fix the ideas, we present a simple example with explicit solution borrowed from [75, Ch. 4, Sec. 4.1.5, p. 106].

Example 2.1 We consider the Laplace problem defined in a ring $B_\rho \setminus \overline{B_\varepsilon}$, where B_ρ is a ball of radius $\rho > \varepsilon$. By taking $b = 0$ and $\partial_n u_\varepsilon = \cos\theta$ on ∂B_ρ, the boundary value problem defined in the topologically perturbed geometrical domain Ω_ε can be stated as: Find u_ε, such that

$$\begin{cases} \Delta u_\varepsilon = 0 & \text{in } B_\rho \setminus \overline{B_\varepsilon}, \\ \partial_n u_\varepsilon = \cos\theta \text{ on } & \partial B_\rho, \\ \partial_n u_\varepsilon = 0 & \text{on } \partial B_\varepsilon. \end{cases} \tag{2.38}$$

The associated explicit solution written in a polar coordinate system (r, θ) with center at the ring is given, up to an arbitrary additive constant, by

$$u_\varepsilon(r, \theta) = \frac{\rho^2}{r} \left(\frac{r^2 + \varepsilon^2}{\rho^2 - \varepsilon^2} \right) \cos\theta. \tag{2.39}$$

Thus, the energy shape function can be evaluated explicitly, namely

$$\psi(\chi_\varepsilon) = -\frac{\pi\rho^2}{2} \left(\frac{\rho^2 + \varepsilon^2}{\rho^2 - \varepsilon^2} \right), \tag{2.40}$$

which can be expanded in power of ε as follows:

$$\psi(\chi_\varepsilon) = -\frac{1}{2}\pi\rho^2 - \pi\varepsilon^2 + O(\varepsilon^4). \tag{2.41}$$

On the other hand, according to (2.37) and by considering $u(r, \theta) = r\cos\theta$, the topological asymptotic expansion of the energy shape function can be written as

$$\psi(\chi_\varepsilon) = \psi(\chi) - \pi\varepsilon^2 \|\nabla u\|^2 + o(\varepsilon^2)$$
$$= -\frac{1}{2}\pi\rho^2 - \pi\varepsilon^2 + o(\varepsilon^2), \tag{2.42}$$

since $\|\nabla u\|^2 = 1$, which corroborates with the above expansion in power of ε. In particular, by setting $\rho = 1$, we can define the following quantity:

$$\delta\psi(\varepsilon) := \frac{\psi(\chi_\varepsilon) - \psi(\chi)}{f(\varepsilon)} = -1 - \frac{\varepsilon^2}{1 - \varepsilon^2}. \tag{2.43}$$

Fig. 2.2 Variation of the energy shape functional evaluated explicitly in the ring for the Neumann case

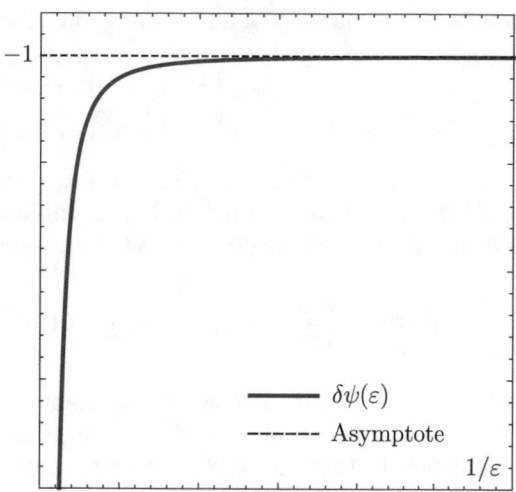

This result is represented in the graph $\delta\psi(\varepsilon) \times 1/\varepsilon$ of Fig. 2.2, where we observe that the horizontal asymptote (dashed line) corresponds to the topological derivative evaluated at the center of the disk, whose value is -1.

2.3.2 Dirichlet Boundary Condition on the Hole

By taking $\beta = 1$ in (2.7), we have homogeneous Dirichlet boundary condition on ∂B_ε, that is $u_\varepsilon|_{\partial B_\varepsilon} = 0 \Rightarrow \partial_\tau u_\varepsilon|_{\partial B_\varepsilon} = 0$, where τ is the unit tangent vector field on ∂B_ε. In this case, the variation of the total potential energy given by (2.19) reduces itself to

$$\mathscr{J}_{\Omega_\varepsilon}(u_\varepsilon) - \mathscr{J}_\Omega(u) = -\frac{1}{2}\int_{\partial B_\varepsilon} u\partial_n u_\varepsilon + \frac{1}{2}\pi\varepsilon^2 bu(\widehat{x}) + o(\varepsilon^2). \tag{2.44}$$

Let us consider the following *ansatz* for an expansion of u_ε [58, 68]

$$u_\varepsilon(x) = u(x) + v_\varepsilon(x) + \varepsilon w_\varepsilon(\varepsilon^{-1}x) + \widetilde{u}_\varepsilon(x)$$

$$= u(\widehat{x}) + \nabla u(\widehat{x}) \cdot (x - \widehat{x}) + \frac{1}{2}\nabla\nabla u(y)(x - \widehat{x}) \cdot (x - \widehat{x})$$

$$+ v_\varepsilon(x) + \varepsilon w_\varepsilon(\varepsilon^{-1}x) + \widetilde{u}_\varepsilon(x), \tag{2.45}$$

where y is used to denote an intermediate point between x and $\widehat{x}$. Moreover, function v_ε is defined as

$$v_\varepsilon(x) = \alpha(\varepsilon)u(\widehat{x})G(x), \tag{2.46}$$

where G is the solution of the following boundary value problem: Find G, such that

$$\begin{cases} -\Delta G = \delta(x - \widehat{x}) & \text{in } \Omega, \\ \quad\quad G = 0 & \text{on } \partial\Omega, \end{cases} \qquad (2.47)$$

with $\delta(x - \widehat{x})$ used to denote the *Dirac mass* concentrated at the point $\widehat{x} \in \Omega$. From the *fundamental solution* for the Laplacian, function G admits a representation in the neighborhood of the point $\widehat{x} \in \Omega$ in the form

$$G(x) = -\left(\frac{1}{2\pi} \log \|x - \widehat{x}\| + g(x)\right), \quad \text{with} \quad \|x - \widehat{x}\| > 0, \qquad (2.48)$$

where g is harmonic in Ω and has to compensate for the discrepancy left by the above representation on $\partial\Omega$. That is, function g is the solution of the following auxiliary boundary value problem: Find g, such that

$$\begin{cases} \Delta g = 0 & \text{in } \Omega, \\ g = -\frac{1}{2\pi} \log \|x - \widehat{x}\| & \text{on } \partial\Omega. \end{cases} \qquad (2.49)$$

Therefore, in the neighborhood of the hole, function v_ε can be written as

$$v_\varepsilon(x) = -\alpha(\varepsilon)u(\widehat{x}) \left(\frac{1}{2\pi} \log \|x - \widehat{x}\| + g(x)\right). \qquad (2.50)$$

On the boundary of the hole ∂B_ε we have $u_\varepsilon|_{\partial B_\varepsilon} = 0$, so that the expansion for u_ε, evaluated on ∂B_ε, yields

$$u(\widehat{x}) - \varepsilon \nabla u(\widehat{x}) \cdot n + \varepsilon^2 \nabla\nabla u(y)n \cdot n$$

$$-\alpha(\varepsilon)u(\widehat{x}) \left(\frac{1}{2\pi} \log \varepsilon + g(\widehat{x}) - \varepsilon \nabla g(\widehat{x}) \cdot n + \varepsilon^2 \nabla\nabla g(z)n \cdot n\right)$$

$$+\varepsilon w_\varepsilon \left(\varepsilon^{-1}x\right) + \widetilde{u}_\varepsilon(x) = 0, \qquad (2.51)$$

where z is an intermediate point between x and $\widehat{x}$. Now, we can construct $\widetilde{u}_\varepsilon$ in such a way that it compensates for the discrepancies left by the higher order terms in ε, namely

$$\widetilde{u}_\varepsilon(x) = \varepsilon^2(\alpha(\varepsilon)u(\widehat{x})\nabla\nabla g(z)n - \nabla\nabla u(y)n) \cdot n \quad \text{on} \quad \partial B_\varepsilon. \qquad (2.52)$$

In the fast variable $\xi = \varepsilon^{-1}x$, function w_ε is the solution of the following *exterior problem*: Find w_ε, such that

$$\begin{cases} \Delta_\xi w_\varepsilon = 0 & \text{in } \mathbb{R}^2 \setminus \overline{B_1}, \\ w_\varepsilon \to 0 & \text{at } \infty, \\ w_\varepsilon = (\nabla u(\widehat{x}) - \alpha(\varepsilon)u(\widehat{x})\nabla g(\widehat{x})) \cdot n & \text{on } \partial B_1, \end{cases} \qquad (2.53)$$

which has an *explicit solution* of the form

$$w_\varepsilon(\varepsilon^{-1}x) = -\frac{\varepsilon}{\|x - \widehat{x}\|^2}(\nabla u(\widehat{x}) - \alpha(\varepsilon)u(\widehat{x})\nabla g(\widehat{x})) \cdot (x - \widehat{x}). \tag{2.54}$$

Thus, the previous expansion given by (2.51) reduces itself to

$$u(\widehat{x}) - \alpha(\varepsilon)u(\widehat{x})\left(\frac{1}{2\pi}\log\varepsilon + g(\widehat{x})\right) = 0 \quad \text{on} \quad \partial B_\varepsilon, \tag{2.55}$$

which can be solved in terms of $\alpha(\varepsilon)$, leading to

$$\alpha(\varepsilon) = \frac{2\pi}{\log\varepsilon + 2\pi g(\widehat{x})}. \tag{2.56}$$

Finally, the *remainder* $\widetilde{u}_\varepsilon$ is constructed in order to compensate for the discrepancies previously introduced in the expansion of u_ε. Thus, $\widetilde{u}_\varepsilon$ has to be the solution of the following boundary value problem: Find $\widetilde{u}_\varepsilon$, such that

$$\begin{cases} \Delta\widetilde{u}_\varepsilon = 0 & \text{in } \Omega_\varepsilon, \\ \widetilde{u}_\varepsilon = -\varepsilon w_\varepsilon & \text{on } \partial\Omega, \\ \widetilde{u}_\varepsilon = \varepsilon^2(\alpha(\varepsilon)u(\widehat{x})\nabla\nabla g(z)n - \nabla\nabla u(y)n) \cdot n & \text{on } \partial B_\varepsilon, \end{cases} \tag{2.57}$$

where y and z are intermediate points between x and $\widehat{x}$. Under appropriate regularity conditions, the estimate $\widetilde{u}_\varepsilon \approx O(\varepsilon^2)$ holds true in a suitable norm, since $w_\varepsilon \approx O(\varepsilon)$ on the exterior boundary $\partial\Omega$. Therefore, the *expansion* for u_ε can be written as

$$u_\varepsilon(x) = u(x) - \alpha(\varepsilon)u(\widehat{x})\left(\frac{1}{2\pi}\log\|x - \widehat{x}\| + g(x)\right)$$

$$-\frac{\varepsilon^?}{\|x - \widehat{x}\|^2}(\nabla u(\widehat{x}) - \alpha(\varepsilon)u(\widehat{x})\nabla g(\widehat{x})) \cdot (x - \widehat{x}) + O(\varepsilon^2), \tag{2.58}$$

with $\alpha(\varepsilon)$ given by (2.56).

Let us introduce the above ansatz into the first term of (2.44) to obtain

$$\int_{\partial B_\varepsilon} u\partial_n u_\varepsilon = \int_{\partial B_\varepsilon} u\partial_n u + \int_{\partial B_\varepsilon} u\partial_n v_\varepsilon + \varepsilon\int_{\partial B_\varepsilon} u\partial_n w_\varepsilon(\varepsilon^{-1}x) + o(\varepsilon^2), \tag{2.59}$$

where, as before, the first term of the above equation admits the following expansion in power of ε

$$\int_{\partial B_\varepsilon} u\partial_n u = -\pi\varepsilon^2\|\nabla u(\widehat{x})\|^2 + \pi\varepsilon^2 bu(\widehat{x}) + o(\varepsilon^2). \tag{2.60}$$

Furthermore, function u, evaluated on ∂B_ε, can be expanded in Taylor series as follows

$$u(x)|_{\partial B_\varepsilon} = u(\widehat{x}) - \varepsilon \nabla u(\widehat{x}) \cdot n + \varepsilon^2 \nabla \nabla u(y) n \cdot n$$
$$= u(\widehat{x}) - \varepsilon \nabla u(\widehat{x}) \cdot n + O(\varepsilon^2), \tag{2.61}$$

where y is an intermediate point between x and $\widehat{x}$. On the other hand, the normal derivative of function v_ε given by (2.50), evaluated on ∂B_ε, can be written as

$$\partial v_\varepsilon(x)|_{\partial B_\varepsilon} = \alpha(\varepsilon) u(\widehat{x}) \left(\frac{1}{2\pi \varepsilon} - \nabla g(\widehat{x}) \cdot n + \varepsilon \nabla \nabla g(z) n \cdot n \right)$$
$$= \alpha(\varepsilon) u(\widehat{x}) \left(\frac{1}{2\pi \varepsilon} - \nabla g(\widehat{x}) \cdot n \right) + o(\varepsilon), \tag{2.62}$$

with z used to denote an intermediate point between x and $\widehat{x}$. Finally, the normal derivative of function $w_\varepsilon(\varepsilon^{-1}x)$ given by (2.54), evaluated on ∂B_ε, can be written as

$$\partial w_\varepsilon(\varepsilon^{-1}x)|_{\partial B_\varepsilon} = \frac{1}{\varepsilon} \left(\nabla u(\widehat{x}) - \alpha(\varepsilon) u(\widehat{x}) \nabla g(\widehat{x}) \right) \cdot n. \tag{2.63}$$

By taking into account these derived results, we observe that

$$\int_{\partial B_\varepsilon} u \partial_n v_\varepsilon = \alpha(\varepsilon) u(\widehat{x}) \int_{\partial B_\varepsilon} (u(\widehat{x}) - \varepsilon \nabla u(\widehat{x}) \cdot n) \left(\frac{1}{2\pi \varepsilon} - \nabla g(\widehat{x}) \cdot n \right) + o(\varepsilon^2). \tag{2.64}$$

After evaluating each one of these integrals by hand, we have

$$\frac{1}{2\pi \varepsilon} \int_{\partial B_\varepsilon} u(\widehat{x}) = u(\widehat{x}), \tag{2.65}$$

$$\frac{1}{2\pi} \int_{\partial B_\varepsilon} \nabla u(\widehat{x}) \cdot n = 0, \tag{2.66}$$

$$u(\widehat{x}) \int_{\partial B_\varepsilon} \nabla g(\widehat{x}) \cdot n = 0, \tag{2.67}$$

$$\varepsilon \int_{\partial B_\varepsilon} (\nabla u(\widehat{x}) \cdot n)(\nabla g(\widehat{x}) \cdot n) = O(\varepsilon^2), \tag{2.68}$$

which results in

$$\int_{\partial B_\varepsilon} u \partial_n v_\varepsilon = \alpha(\varepsilon)|u(\widehat{x})|^2 + o(\varepsilon^2), \tag{2.69}$$

since $\alpha(\varepsilon) \times O(\varepsilon^2) = o(\varepsilon^2)$. In the same way,

$$\varepsilon \int_{\partial B_\varepsilon} u \partial_n w_\varepsilon(\varepsilon^{-1}x) = \int_{\partial B_\varepsilon} (u(\widehat{x}) - \varepsilon \nabla u(\widehat{x}) \cdot n) (\nabla u(\widehat{x}) - \alpha(\varepsilon)u(\widehat{x})\nabla g(\widehat{x})) \cdot n$$

$$+ o(\varepsilon^2), \tag{2.70}$$

where, by evaluating again each one of these integrals by hand, we have

$$u(\widehat{x}) \int_{\partial B_\varepsilon} \nabla u(\widehat{x}) \cdot n = 0, \tag{2.71}$$

$$\varepsilon \int_{\partial B_\varepsilon} (\nabla u(\widehat{x}) \cdot n) (\nabla u(\widehat{x}) \cdot n) = \pi \varepsilon^2 \|\nabla u(\widehat{x})\|^2, \tag{2.72}$$

$$\alpha(\varepsilon)|u(\widehat{x})|^2 \int_{\partial B_\varepsilon} \nabla g(\widehat{x}) \cdot n = 0, \tag{2.73}$$

$$\varepsilon \alpha(\varepsilon)u(\widehat{x}) \int_{\partial B_\varepsilon} (\nabla u(\widehat{x}) \cdot n)(\nabla u(\widehat{x}) \cdot n) = o(\varepsilon^2), \tag{2.74}$$

where we have used the fact that $\varepsilon^2 \alpha(\varepsilon) = o(\varepsilon^2)$. Therefore,

$$\varepsilon \int_{\partial B_\varepsilon} u \partial_n w_\varepsilon(\varepsilon^{-1}x) = -\pi \varepsilon^2 \|\nabla u(\widehat{x})\|^2 + o(\varepsilon^2). \tag{2.75}$$

After collecting the main results derived here, we conclude that the first term in (2.44) can be expanded in power of ε as follows:

$$\int_{\partial B_\varepsilon} u \partial_n u_\varepsilon = \alpha(\varepsilon)|u(\widehat{x})|^2 - 2\pi \varepsilon^2 \|\nabla u(\widehat{x})\|^2 + \pi \varepsilon^2 bu(\widehat{x}) + o(\varepsilon^2). \tag{2.76}$$

Thus, the above result combined with expansion (2.44) allows for writing the variation of the total potential energy in the following form:

$$\mathscr{J}_{\Omega_\varepsilon}(u_\varepsilon) - \mathscr{J}_\Omega(u) = -\frac{\pi}{\log \varepsilon + 2\pi g(\widehat{x})}|u(\widehat{x})|^2 + \pi \varepsilon^2 \|\nabla u(\widehat{x})\|^2 + o(\varepsilon^2)$$

$$= -\frac{\pi}{\log \varepsilon + 2\pi g(\widehat{x})}|u(\widehat{x})|^2 + O(\varepsilon^2). \tag{2.77}$$

Now, in order to identify the leading term in the above expansion, we chose

$$f(\varepsilon) = -\frac{\pi}{\log \varepsilon + 2\pi g(\widehat{x})}, \tag{2.78}$$

which leads to the final formula for the *topological derivative*, namely [31, 76]

$$\mathscr{T}(\widehat{x}) = |u(\widehat{x})|^2 \quad \forall\, \widehat{x} \in \Omega. \tag{2.79}$$

Finally, the *topological asymptotic expansion* of the energy shape function can be written as

$$\psi(\chi_\varepsilon(\widehat{x})) = \psi(\chi) - \frac{\pi}{\log \varepsilon + 2\pi g(\widehat{x})} |u(\widehat{x})|^2 + O(\varepsilon^2). \tag{2.80}$$

The complete mathematical justification for the above expansion can be found in the book by Novotny and Sokolowski [75, Ch. 10], for instance.

Remark 2.1 It is important to observe that the approximation given by

$$f(\varepsilon) \approx -\pi/\log \varepsilon \tag{2.81}$$

can frequently be found in the literature, leading to the following simplified expansion (see, for instance, [50])

$$\psi(\chi_\varepsilon(\widehat{x})) = \psi(\chi) - \frac{\pi}{\log \varepsilon} |u(\widehat{x})|^2 + o\left(\frac{-1}{\log \varepsilon}\right), \tag{2.82}$$

which introduces a discrepancy on the resulting topological asymptotic expansion.

In addition, we can go further in the expansion. In fact, let us consider one more term is the topological asymptotic expansion of the form

$$\psi(\chi_\varepsilon(\widehat{x})) = \psi(\chi) + f(\varepsilon)\mathscr{T}(\widehat{x}) + f_2(\varepsilon)\mathscr{T}^2(\widehat{x}) + o(f_2(\varepsilon)), \tag{2.83}$$

where $f_2(\varepsilon)$ is a second order correction function which decays monotonically such that $f_2(\varepsilon) \to 0$ with $\varepsilon \to 0$. Furthermore,

$$\lim_{\varepsilon \to 0} \frac{f_2(\varepsilon)}{f(\varepsilon)} = 0. \tag{2.84}$$

Thus, $\mathscr{T}$ and $\mathscr{T}^2$ are identified as the first and second order topological derivatives of the shape functional ψ, respectively. After dividing (2.83) by $f_2(\varepsilon)$, the limit passage $\varepsilon \to 0$ allows for defining the *second order topological* as follows:

$$\mathscr{T}^2(\widehat{x}) := \lim_{\varepsilon \to 0} \frac{\psi(\chi_\varepsilon(\widehat{x})) - \psi(\chi) - f(\varepsilon)\mathscr{T}(\widehat{x})}{f_2(\varepsilon)}. \tag{2.85}$$

From these elements, it is possible to evaluate the second order topological derivative of the energy shape functional from the obtained expansion. In fact, by combining (2.77), first line, with (2.78) and (2.79), we have

$$\psi(\chi_\varepsilon(\widehat{x})) - \psi(\chi) - f(\varepsilon)\mathscr{T}(\widehat{x}) = \pi\varepsilon^2\|\nabla u(\widehat{x})\|^2 + o(\varepsilon^2). \qquad (2.86)$$

In order to identify the leading term of the above expansion, we choose

$$f_2(\varepsilon) = \pi\varepsilon^2, \qquad (2.87)$$

which leads to the final formula for the *second order topological derivative* [83], namely

$$\mathscr{T}^2(\widehat{x}) = \|\nabla u(\widehat{x})\|^2 \quad \forall \widehat{x} \in \Omega. \qquad (2.88)$$

Finally, the *topological asymptotic expansion* of the energy shape functional takes the form

$$\psi(\chi_\varepsilon(\widehat{x})) = \psi(\chi) - \frac{\pi}{\log\varepsilon + 2\pi g(\widehat{x})}|u(\widehat{x})|^2 + \pi\varepsilon^2\|\nabla u(\widehat{x})\|^2 + o(\varepsilon^2). \qquad (2.89)$$

In order to fix the ideas, let us present a simple example with explicit solution borrowed from [75, Ch. 4, Sec. 4.1.5, p. 107].

Example 2.2 Let us consider again the Laplace problem defined in a ring $B_\rho \setminus \overline{B_\varepsilon}$, where B_ρ is a ball of radius $\rho > \varepsilon$. By taking $b = 0$ and $u_\varepsilon = a + \cos\theta$ on ∂B_ρ, the boundary value problem associated with the topologically perturbed domain Ω_ε reads: Find u_ε, such that

$$\begin{cases} \Delta u_\varepsilon = 0 & \text{in } B_\rho \setminus \overline{B_\varepsilon}, \\ u_\varepsilon = a + \cos\theta & \text{on } \partial B_\rho, \\ u_\varepsilon = 0 & \text{on } \partial B_\varepsilon, \end{cases} \qquad (2.90)$$

whose explicit solution, written in a polar coordinate system (r, θ) with center at the ring, is given by

$$u_\varepsilon(r, \theta) = a\frac{\log(r/\varepsilon)}{\log(\rho/\varepsilon)} + \frac{\rho}{r}\left(\frac{r^2 - \varepsilon^2}{\rho^2 - \varepsilon^2}\right)\cos\theta. \qquad (2.91)$$

Thus, the shape functional can be evaluated explicitly as follows:

$$\psi(\chi_\varepsilon) = \frac{\pi}{\log(\rho/\varepsilon)}a^2 + \frac{\pi}{2}\frac{\rho^2 + \varepsilon^2}{\rho^2 - \varepsilon^2}. \qquad (2.92)$$

After expanding it in power of ε, we obtain

$$\psi(\chi_\varepsilon) = \frac{\pi}{2} + \frac{\pi}{\log(\rho/\varepsilon)}a^2 + \pi\varepsilon^2\frac{1}{\rho^2} + O(\varepsilon^4). \qquad (2.93)$$

On the other hand, since $u(r, \theta) = a + (r/\rho)\cos\theta$, the topological asymptotic expansion (2.80) reads

$$\psi(\chi_\varepsilon) = \psi(\chi) - \frac{\pi}{\log\varepsilon + 2\pi g}|u|^2 + \pi\varepsilon^2\|\nabla u\|^2 + o(\varepsilon^2)$$

$$= \frac{\pi}{2} + \frac{\pi}{\log(\rho/\varepsilon)}a^2 + \pi\varepsilon^2\frac{1}{\rho^2} + o(\varepsilon^2), \tag{2.94}$$

which corroborates with the above expansion in power of ε, provided that $\|\nabla u\|^2 = 1/\rho^2$, where g is the solution of (2.49), that is

$$\begin{cases} \Delta g = 0 & \text{in } B_\rho \\ g = -\frac{1}{2\pi}\log\rho & \text{on } \partial B_\rho \end{cases} \Rightarrow g(\widehat{x}) = -\frac{1}{2\pi}\log\rho. \tag{2.95}$$

By choosing $\rho = a = 1$, we can introduce the following quantities:

$$\delta\psi_1(\varepsilon) := \frac{\psi(\chi_\varepsilon) - \psi(\chi)}{f(\varepsilon)} = 1 - \frac{\varepsilon^2}{1 - \varepsilon^2}\log\varepsilon, \tag{2.96}$$

$$\delta\psi_2(\varepsilon) := \frac{\psi(\chi_\varepsilon) - \psi(\chi) - f(\varepsilon)\mathscr{T}(\widehat{x})}{f_2(\varepsilon)} = 1 + \frac{\varepsilon^2}{1 - \varepsilon^2}. \tag{2.97}$$

These two results are represented in the graphs $\delta\psi_1(\varepsilon) \times 1/\varepsilon$ (blue line) and $\delta\psi_2(\varepsilon) \times 1/\varepsilon$ (red line) of Fig. 2.3. We observe that the horizontal asymptote (dashed line) associated with $\delta\psi_1(\varepsilon)$ and $\delta\psi_2(\varepsilon)$ corresponds to the first and second order topological derivatives, respectively, evaluated at the center of the disk, whose value is $+1$ for both cases.

Fig. 2.3 Variation of the energy shape functional evaluated explicitly in the ring for the Dirichlet case

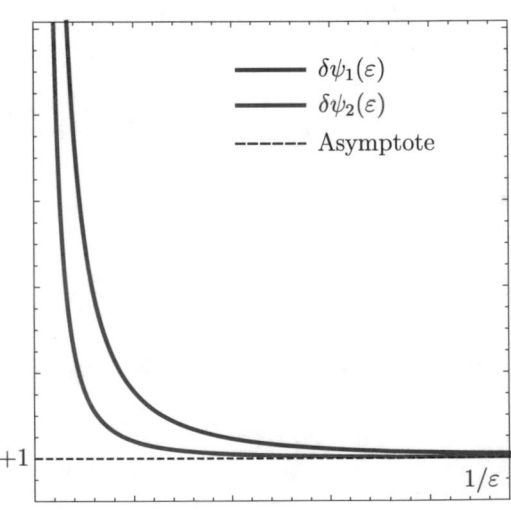

2.4   Summary of the Results

In this chapter we have evaluated the topological derivatives for the energy shape functional associated with the Poisson's equation into two-spatial dimensions, by taking into account homogeneous Neumann or Dirichlet conditions on the boundary of the hole.

For the sake of completeness, let us also consider Robin condition on the boundary of the hole. In this case, we have to add one more term to the total potential energy (2.4), namely

$$\psi(\chi_\varepsilon) := \mathscr{J}_{\Omega_\varepsilon}(u_\varepsilon) = \frac{1}{2}\int_{\Omega_\varepsilon} \|\nabla u_\varepsilon\|^2 + \frac{1}{2}\int_{\partial B_\varepsilon} |u_\varepsilon|^2 - \int_{\Omega_\varepsilon} bu_\varepsilon, \qquad (2.98)$$

where u_ε is the solution to the following boundary value problem endowed with Robin condition on the boundary of the hole: Find u_ε, such that

$$\begin{cases} -\Delta u_\varepsilon = b \ \text{in} \ \Omega_\varepsilon, \\ u_\varepsilon = 0 \ \text{on} \ \partial\Omega, \\ \partial_n u_\varepsilon + u_\varepsilon = 0 \ \text{on} \ \partial B_\varepsilon. \end{cases} \qquad (2.99)$$

The solution to the above boundary value problem admits the following *expansion*

$$u_\varepsilon(x) = u(x) + \tilde{u}_\varepsilon(x), \qquad (2.100)$$

where $\tilde{u}_\varepsilon$ is the remainder. Therefore, the variation of the energy shape functional given by (2.19), taking into account the new term regarding Robin boundary condition on the hole, can be written as

$$\mathscr{J}_{\Omega_\varepsilon}(u_\varepsilon) - \mathscr{J}_{\Omega}(u) = \frac{1}{2}\int_{\partial B_\varepsilon} (u_\varepsilon \partial_n u - u \partial_n u_\varepsilon) + o(\varepsilon)$$

$$= \frac{1}{2}\int_{\partial B_\varepsilon} (\partial_n u + u)u_\varepsilon + o(\varepsilon) = \frac{1}{2}\int_{\partial B_\varepsilon} u_\varepsilon u + o(\varepsilon), \quad (2.101)$$

since $\partial_n u_\varepsilon = -u_\varepsilon$ on ∂B_ε. After replacing (2.100) in the above expansion, we can evaluate the resulting integral on the boundary of hole ∂B_ε explicitly, namely

$$\mathscr{J}_{\Omega_\varepsilon}(u_\varepsilon) - \mathscr{J}_{\Omega}(u) = \pi\varepsilon \, |u(\widehat{x})|^2 + o(\varepsilon). \qquad (2.102)$$

Now, in order to identify the leading term of the above expansion, we choose

$$f(\varepsilon) = \pi\varepsilon, \qquad (2.103)$$

which leads to the final formula for the *topological derivative*, namely [76]

$$\mathscr{T}(\widehat{x}) = |u(\widehat{x})|^2 \quad \forall \widehat{x} \in \Omega. \qquad (2.104)$$

Table 2.1 Topological derivatives of the total potential energy associated with the Poisson's problem into two spatial dimensions, taking into account homogeneous Neumann, Dirichlet, or Robin boundary conditions on the hole

Boundary condition	$f(\varepsilon)$	$\mathscr{T}(\widehat{x})$
Neumann	$\pi\varepsilon^2$	$-\|\nabla u(\widehat{x})\|^2 + bu(\widehat{x})$
Dirichlet	$-\dfrac{\pi}{\log\varepsilon + 2\pi g(\widehat{x})}$	$|u(\widehat{x})|^2$
Robin	$\pi\varepsilon$	$|u(\widehat{x})|^2$

Finally, the *topological asymptotic expansion* of the energy shape functional takes the form

$$\psi(\chi_\varepsilon(\widehat{x})) = \psi(\chi) + \pi\varepsilon\,|u(\widehat{x})|^2 + o(\varepsilon). \tag{2.105}$$

Let us now summarize the results for *topological derivatives* obtained in this chapter, which are reported in Table 2.1. Recalling that u is the solution to (2.3) and g is the solution to (2.49), both defined in the unperturbed domain Ω.

2.5 Exercises

1. From (2.4), derive (2.5) and (2.7).
2. By using separation of variables, find the explicit solutions to the boundary value problems (2.24) and (2.53).
3. Consider the problem defined in a ring of internal radius equal to ε and external radius equal to one: Find u_ε, such that

$$\begin{cases} \Delta u_\varepsilon = 0 & \text{in } B_1 \setminus \overline{B_\varepsilon}, \\ u_\varepsilon = a + \cos\theta & \text{on } \partial B_1, \\ u_\varepsilon + \partial_n u_\varepsilon = 0 & \text{on } \partial B_\varepsilon. \end{cases}$$

By taking into account a shape functional of the form

$$\psi(\chi_\varepsilon) = \frac{1}{2}\int_{B_1\setminus\overline{B_\varepsilon}} \|\nabla u_\varepsilon\|^2 + \frac{1}{2}\int_{\partial B_c} |u_\varepsilon|^2,$$

develop $\psi(\chi_\varepsilon)$ in power of ε around the origin to obtain

$$\psi(\chi_\varepsilon) = \frac{\pi}{2} + \pi\varepsilon\,a^2 + o(\varepsilon),$$

and compare it with the topological asymptotic expansion (2.105).

Chapter 3
Regular Domain Perturbation

In this chapter we deal with the topological derivative of the so-called compliance shape functional associated with a modified Helmholtz problem, with respect to the nucleation of a small inclusion represented by B_ε. Similar analysis can be found in the paper by Amstutz [9], for instance. In particular, the topologically perturbed domain is obtained after nucleating a circular hole $B_\varepsilon(\widehat{x})$ within $\Omega \subset \mathbb{R}^2$, where $\overline{B_\varepsilon(\widehat{x})} \subset \Omega$ denotes a ball of radius ε and center at $\widehat{x} \in \Omega$. Then, the hole produced by $B_\varepsilon(\widehat{x})$ is filled by an inclusion with different material property from the background, as shown in Fig. 3.1. The material properties are characterized by a piecewise constant function γ_ε, assuming the value 1 in $\Omega \setminus \overline{B_\varepsilon}$ and γ in B_ε, where γ is the *contrast*.

3.1 Problem Formulation

The compliance *shape functional* associated with the unperturbed domain is defined as

$$\psi(\chi) := \mathscr{J}(u) = \int_\Omega bu , \qquad (3.1)$$

where b is a given source term assumed to be smooth enough and the scalar function $u : \Omega \mapsto \mathbb{R}$ is the solution of the following variational problem:

$$u \in H_0^1(\Omega) : \int_\Omega \nabla u \cdot \nabla \eta + \int_\Omega u\eta = \int_\Omega b\eta \quad \forall \eta \in H_0^1(\Omega) . \qquad (3.2)$$

The strong equation associated with the variational problem (3.2) is given by: Find u, such that

© The Author(s), under exclusive license to Springer Nature Switzerland AG 2020
A. A. Novotny, J. Sokołowski, *An Introduction to the Topological
Derivative Method*, SpringerBriefs in Mathematics,
https://doi.org/10.1007/978-3-030-36915-6_3

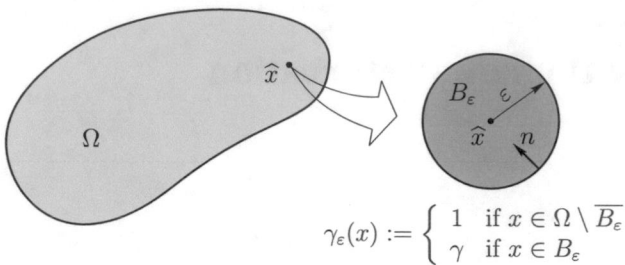

$$\gamma_\varepsilon(x) := \begin{cases} 1 & \text{if } x \in \Omega \setminus \overline{B_\varepsilon} \\ \gamma & \text{if } x \in B_\varepsilon \end{cases}$$

Fig. 3.1 Topologically perturbed domain by the nucleation of a small circular inclusion

$$\begin{cases} -\Delta u + u = b \ \text{ in } \Omega \ , \\ \qquad\quad u = 0 \ \text{on } \partial\Omega \ . \end{cases} \tag{3.3}$$

Now, let us consider the following piecewise constant functions:

$$\alpha_\varepsilon(x) := \begin{cases} 1 & \text{if } x \in \Omega \setminus \overline{B_\varepsilon} \\ \alpha & \text{if } x \in B_\varepsilon, \end{cases} \tag{3.4}$$

$$\beta_\varepsilon(x) := \begin{cases} 1 & \text{if } x \in \Omega \setminus \overline{B_\varepsilon} \\ \beta & \text{if } x \in B_\varepsilon, \end{cases} \tag{3.5}$$

$$\gamma_\varepsilon(x) := \begin{cases} 1 & \text{if } x \in \Omega \setminus \overline{B_\varepsilon} \\ \gamma & \text{if } x \in B_\varepsilon, \end{cases} \tag{3.6}$$

where $\alpha, \beta \in \mathbb{R}^+$ and $\gamma \in \mathbb{R}$ are the *contrasts* into the material properties. From these elements we can introduce the topologically perturbed counterpart of the problem. In particular, the compliance shape functional associated with the perturbed domain is given by

$$\psi(\chi_\varepsilon) := \mathscr{J}_\varepsilon(u_\varepsilon) = \int_\Omega \gamma_\varepsilon \, b u_\varepsilon \ , \tag{3.7}$$

where the scalar function $u_\varepsilon : \Omega \mapsto \mathbb{R}$ is the solution of the following variational problem:

$$u_\varepsilon \in H_0^1(\Omega) \ : \ \int_\Omega \alpha_\varepsilon \nabla u_\varepsilon \cdot \nabla \eta + \int_\Omega \beta_\varepsilon \, u_\varepsilon \eta = \int_\Omega \gamma_\varepsilon \, b\eta \quad \forall \eta \in H_0^1(\Omega) \ . \tag{3.8}$$

The *strong equation* associated with the variational problem (3.8) can be written as: Find u_ε, such that

$$
\begin{cases}
-\operatorname{div}(\alpha_\varepsilon \nabla u_\varepsilon) + \beta_\varepsilon u_\varepsilon = \gamma_\varepsilon b & \text{in } \Omega\,, \\
u_\varepsilon = 0 & \text{on } \partial\Omega\,, \\
\left.\begin{aligned} u_\varepsilon|_{\Omega\setminus\overline{B_\varepsilon}} - u_\varepsilon|_{B_\varepsilon} = 0 \\ \partial_n u_\varepsilon|_{\Omega\setminus\overline{B_\varepsilon}} - \alpha\partial_n u_\varepsilon|_{B_\varepsilon} = 0 \end{aligned}\right\} & \text{on } \partial B_\varepsilon\,.
\end{cases}
\tag{3.9}
$$

The *transmission condition* on the boundary of the inclusion ∂B_ε are obtained naturally from the variational formulation (3.8).

3.2 Existence of the Topological Derivative

The existence of the associated topological derivative is ensured by the following lemma:

Lemma 3.1 *Let u and u_ε be the solutions of the original (3.2) and perturbed (3.8) problems, respectively. Then, the following estimate holds true:*

$$
\|u_\varepsilon - u\|_{H^1(\Omega)} \le C\varepsilon\,,
\tag{3.10}
$$

where C is a constant independent of the small parameter ε.

Proof From the definition of the contrasts given by (3.4)–(3.6), we have

$$
\begin{aligned}
\int_\Omega \nabla u \cdot \nabla \eta &= \int_{\Omega\setminus\overline{B_\varepsilon}} \nabla u \cdot \nabla \eta + \int_{B_\varepsilon} \nabla u \cdot \nabla \eta \pm \int_{B_\varepsilon} \alpha \nabla u \cdot \nabla \eta \\
&= \int_\Omega \alpha_\varepsilon \nabla u \cdot \nabla \eta + (1-\alpha)\int_{B_\varepsilon} \nabla u \cdot \nabla \eta\,,
\end{aligned}
\tag{3.11}
$$

$$
\int_\Omega u\eta = \int_{\Omega\setminus\overline{B_\varepsilon}} u\eta + \int_{B_\varepsilon} u\eta \pm \int_{B_\varepsilon} \beta\, u\eta = \int_\Omega \beta_\varepsilon\, u\eta + (1-\beta)\int_{B_\varepsilon} u\eta\,,
\tag{3.12}
$$

$$
\int_\Omega b\eta = \int_{\Omega\setminus\overline{B_\varepsilon}} b\eta + \int_{B_\varepsilon} b\eta \pm \int_{B_\varepsilon} \gamma\, b\eta = \int_\Omega \gamma_\varepsilon\, b\eta + (1-\gamma)\int_{B_\varepsilon} b\eta\,.
\tag{3.13}
$$

Therefore, Eq. (3.2) can be rewritten as

$$
\begin{aligned}
&\int_\Omega \alpha_\varepsilon \nabla u \cdot \nabla \eta + \int_\Omega \beta_\varepsilon\, u\eta \\
&= \int_\Omega \gamma_\varepsilon\, b\eta - (1-\alpha)\int_{B_\varepsilon} \nabla u \cdot \nabla \eta - (1-\beta)\int_{B_\varepsilon} u\eta + (1-\gamma)\int_{B_\varepsilon} b\eta\,.
\end{aligned}
\tag{3.14}
$$

After subtraction the above result from (3.8), we obtain

$$\int_\Omega \alpha_\varepsilon \nabla(u_\varepsilon - u) \cdot \nabla\eta + \int_\Omega \beta_\varepsilon (u_\varepsilon - u)\eta$$

$$= (1-\alpha) \int_{B_\varepsilon} \nabla u \cdot \nabla\eta + (1-\beta) \int_{B_\varepsilon} u\eta - (1-\gamma) \int_{B_\varepsilon} b\eta . \qquad (3.15)$$

Now, by taking $\eta = u_\varepsilon - u$ as test function we obtain the following equality:

$$\int_\Omega \alpha_\varepsilon \|\nabla(u_\varepsilon - u)\|^2 + \int_\Omega \beta_\varepsilon |u_\varepsilon - u|^2$$

$$= (1-\alpha) \int_{B_\varepsilon} \nabla u \cdot \nabla(u_\varepsilon - u) + (1-\beta) \int_{B_\varepsilon} u(u_\varepsilon - u) - (1-\gamma) \int_{B_\varepsilon} b(u_\varepsilon - u) .$$

$$(3.16)$$

From the Cauchy-Schwarz inequality there is

$$\int_\Omega \alpha_\varepsilon \|\nabla(u_\varepsilon - u)\|^2 + \int_\Omega \beta_\varepsilon |u_\varepsilon - u|^2 \le C_1 \|\nabla u\|_{L^2(B_\varepsilon)} \|\nabla(u_\varepsilon - u)\|_{L^2(B_\varepsilon)}$$

$$+ C_2 \|u\|_{L^2(B_\varepsilon)} \|u_\varepsilon - u\|_{L^2(B_\varepsilon)} + C_3 \|b\|_{L^2(B_\varepsilon)} \|u_\varepsilon - u\|_{L^2(B_\varepsilon)}. \qquad (3.17)$$

By taking into account the interior elliptic regularity of function u, it follows that

$$\int_\Omega \alpha_\varepsilon \|\nabla(u_\varepsilon - u)\|^2 + \int_\Omega \beta_\varepsilon |u_\varepsilon - u|^2$$

$$\le C_4\varepsilon \|\nabla(u_\varepsilon - u)\|_{L^2(B_\varepsilon)} + C_5\varepsilon \|u_\varepsilon - u\|_{L^2(B_\varepsilon)}$$

$$\le C_4\varepsilon \|\nabla(u_\varepsilon - u)\|_{L^2(\Omega)} + C_5\varepsilon \|u_\varepsilon - u\|_{L^2(\Omega)}$$

$$\le C_6\varepsilon \|u_\varepsilon - u\|_{H^1(\Omega)}. \qquad (3.18)$$

Finally, from the *coercivity* of the bilinear form on the left-hand side of the above inequality, namely

$$c\|u_\varepsilon - u\|^2_{H^1(\Omega)} \le \int_\Omega \alpha_\varepsilon \|\nabla(u_\varepsilon - u)\|^2 + \int_\Omega \beta_\varepsilon |u_\varepsilon - u|^2 , \qquad (3.19)$$

we obtain

$$\|u_\varepsilon - u\|^2_{H^1(\Omega)} \le C\varepsilon \|u_\varepsilon - u\|_{H^1(\Omega)} , \qquad (3.20)$$

which leads to the result with constant $C = C_6/c$.   □

3.3   Variation of the Compliance Shape Functional

Since the energy-norm is equivalent to the H^1-norm, Lemma 3.1 ensures the existence of the topological derivative of the compliance shape functional in particular and any energy-based shape functional in general. Now, we have to evaluate the variation of the compliance shape functional, namely

$$\mathscr{J}_\varepsilon(u_\varepsilon) - \mathscr{J}(u) = \int_\Omega \gamma_\varepsilon \, bu_\varepsilon - \int_\Omega bu \,. \tag{3.21}$$

From a simple manipulation we can write the above variation in terms of integrals concentrated in the ball B_ε. In fact, we start by using the definition for the contrast given by (3.6) to obtain

$$\mathscr{J}_\varepsilon(u_\varepsilon) - \mathscr{J}(u) = \int_{\Omega \setminus \overline{B_\varepsilon}} bu_\varepsilon + \int_{B_\varepsilon} \gamma \, bu_\varepsilon \pm \int_{B_\varepsilon} bu_\varepsilon - \int_\Omega bu$$

$$= \int_\Omega b(u_\varepsilon - u) - (1 - \gamma) \int_{B_\varepsilon} bu_\varepsilon \,. \tag{3.22}$$

Following the same steps as before, we have

$$\int_\Omega \alpha_\varepsilon \nabla u_\varepsilon \cdot \nabla \eta = \int_\Omega \nabla u_\varepsilon \cdot \nabla \eta - (1 - \alpha) \int_{B_\varepsilon} \nabla u_\varepsilon \cdot \nabla \eta \,, \tag{3.23}$$

$$\int_\Omega \beta_\varepsilon \, u_\varepsilon \eta = \int_\Omega u_\varepsilon \eta - (1 - \beta) \int_{B_\varepsilon} u_\varepsilon \eta \,, \tag{3.24}$$

$$\int_\Omega \gamma_\varepsilon \, b\eta = \int_\Omega b\eta - (1 - \gamma) \int_{B_\varepsilon} b\eta \,, \tag{3.25}$$

where we have used again the contrasts (3.4)–(3.6). Therefore, Eq. (3.8) can be rewritten as

$$\int_\Omega \nabla u_\varepsilon \cdot \nabla \eta + \int_\Omega u_\varepsilon \eta = \int_\Omega b\eta$$

$$+ (1 - \alpha) \int_{B_\varepsilon} \nabla u_\varepsilon \cdot \nabla \eta + (1 - \beta) \int_{B_\varepsilon} u_\varepsilon \eta - (1 - \gamma) \int_{B_\varepsilon} b\eta \,. \tag{3.26}$$

Now, let us set $\eta = u$ as test function in the above equation, then

$$\int_\Omega \nabla u_\varepsilon \cdot \nabla u + \int_\Omega u_\varepsilon u = \int_\Omega bu$$

$$+ (1 - \alpha) \int_{B_\varepsilon} \nabla u_\varepsilon \cdot \nabla u + (1 - \beta) \int_{B_\varepsilon} u_\varepsilon u - (1 - \gamma) \int_{B_\varepsilon} bu \,.$$

$$\tag{3.27}$$

By setting $\eta = u_\varepsilon$ as test functions in (3.2), there is

$$\int_\Omega \nabla u \cdot \nabla u_\varepsilon + \int_\Omega u u_\varepsilon = \int_\Omega b u_\varepsilon . \tag{3.28}$$

After comparing both the equalities, we observe that

$$\int_\Omega b(u_\varepsilon - u) = (1-\alpha)\int_{B_\varepsilon} \nabla u_\varepsilon \cdot \nabla u + (1-\beta)\int_{B_\varepsilon} u_\varepsilon u - (1-\gamma)\int_{B_\varepsilon} bu. \tag{3.29}$$

Therefore, as expected, the variation of the compliance shape function (3.22) leads to integrals concentrated in the ball B_ε, namely

$$\mathscr{J}_\varepsilon(u_\varepsilon) - \mathscr{J}(u) = (1-\alpha)\int_{B_\varepsilon} \nabla u_\varepsilon \cdot \nabla u$$

$$+ (1-\beta)\int_{B_\varepsilon} u_\varepsilon u - (1-\gamma)\int_{B_\varepsilon} b(u_\varepsilon + u). \tag{3.30}$$

3.4   Topological Derivative Evaluation

The variation of the compliance shape functional is exclusively given in terms of integrals concentrated in the inclusion represented by B_ε. Therefore, in order to obtain a topological asymptotic expansion of the form (1.2), we have to know the asymptotic behavior of the solution u_ε with respect to the small parameter ε in the neighborhood of the ball B_ε. In particular, once such an asymptotic behavior is known explicitly, it is possible to identify function $f(\varepsilon)$, which allows for evaluating the limit $\varepsilon \to 0$ in (1.4), leading to the final formula of the topological derivative $\mathscr{T}$ associated with the shape functional ψ. For the sake of presentation, we divide the analysis into three different cases according to the choice of the contrasts α, β, and γ in (3.4), (3.5), and (3.6), respectively. First, the contrasts α and β are freezing, that is, we set them as $\alpha = \beta = 1$, allowing for performing the analysis by taking into account only the contrast $\gamma \neq 1$. Then, we set $\alpha = \gamma = 1$ and do the analysis for $\beta \neq 1$. Finally, we take $\beta = \gamma = 1$ and focus the attention to the case $\alpha \neq 1$, which is much more involved, since α is associated with a topological perturbation in the principal part of the operator.

3.4.1   Perturbation on the Right-Hand Side

By setting $\alpha = \beta = 1$ in the variation of the compliance shape function (3.30), we have

$$\mathscr{J}_\varepsilon(u_\varepsilon) - \mathscr{J}(u) = -(1-\gamma) \int_{B_\varepsilon} b(u_\varepsilon + u) \, . \tag{3.31}$$

Now, let us sum and subtract the term

$$- (1-\gamma) \int_{B_\varepsilon} bu \, , \tag{3.32}$$

which allows for rewriting (3.31) in the following way:

$$\mathscr{J}_\varepsilon(u_\varepsilon) - \mathscr{J}(u) = -2(1-\gamma) \int_{B_\varepsilon} bu + \mathscr{E}_1(\varepsilon) \, . \tag{3.33}$$

The remainder $\mathscr{E}_1(\varepsilon)$ is defined as

$$\mathscr{E}_1(\varepsilon) = (\gamma - 1) \int_{B_\varepsilon} b(u_\varepsilon - u). \tag{3.34}$$

Thanks to the Cauchy-Schwarz inequality and Lemma 1.1, the remainder $\mathscr{E}_1(\varepsilon)$ can be bounded as follows:

$$\begin{aligned}
|\mathscr{E}_1(\varepsilon)| &\leq C_1 \|b\|_{L^2(B_\varepsilon)} \|u_\varepsilon - u\|_{L^2(B_\varepsilon)} \\
&\leq C_2\varepsilon \|u_\varepsilon - u\|_{L^2(B_\varepsilon)} \\
&\leq C_3\varepsilon^{1+\delta} \|u_\varepsilon - u\|_{H^1(\Omega)} \\
&\leq C_4\varepsilon^{2+\delta} = o(\varepsilon^2),
\end{aligned} \tag{3.35}$$

since $0 < \delta < 1$, where we have also used Lemma 3.1. Let us come back to *expansion* (3.33), which can be rewritten as

$$\mathscr{J}_\varepsilon(u_\varepsilon) - \mathscr{J}(u) = -2\pi\varepsilon^2(1-\gamma)b(\widehat{x})u(\widehat{x}) + \sum_{i=1}^{2} \mathscr{E}_i(\varepsilon) \, . \tag{3.36}$$

The remainder $\mathscr{E}_2(\varepsilon)$ is defined as

$$\mathscr{E}_2(\varepsilon) = 2(\gamma - 1) \int_{B_\varepsilon} (bu - b(\widehat{x})u(\widehat{x})) \, , \tag{3.37}$$

which is trivially bounded as follows:

$$|\mathscr{E}_2(\varepsilon)| \leq C\varepsilon^3 = O(\varepsilon^3) \, , \tag{3.38}$$

where we have used the Cauchy-Schwarz inequality and the interior elliptic regularity of function u. Finally, from the expansion (3.36) we can identify $f_\gamma(\varepsilon) := f(\varepsilon) = \pi\varepsilon^2$ and the topological derivative as

$$\mathcal{T}_\gamma(\widehat{x}) = -2(1 - \gamma)b(\widehat{x})u(\widehat{x}) \quad \forall\, \widehat{x} \in \Omega \,, \tag{3.39}$$

provided that the remainders $\mathcal{E}_1(\varepsilon)$ and $\mathcal{E}_2(\varepsilon)$ are of order $o(\varepsilon^2)$, as shown through the estimates (3.35) and (3.38), respectively.

3.4.2 Perturbation on the Lower Order Term

Now, let us set $\alpha = \gamma = 1$ in the variation of the compliance shape function (3.30), so that

$$\mathscr{J}_\varepsilon(u_\varepsilon) - \mathscr{J}(u) = (1 - \beta)\int_{B_\varepsilon} u_\varepsilon u \,. \tag{3.40}$$

After summing and subtracting the term

$$(1 - \beta)\int_{B_\varepsilon} |u|^2 \,, \tag{3.41}$$

the above expression, i.e., (3.40), can be rearranged as follows:

$$\mathscr{J}_\varepsilon(u_\varepsilon) - \mathscr{J}(u) = (1 - \beta)\int_{B_\varepsilon} |u|^2 + \mathcal{E}_3(\varepsilon) \,. \tag{3.42}$$

The remainder $\mathcal{E}_3(\varepsilon)$ is defined as

$$\mathcal{E}_3(\varepsilon) = (1 - \beta)\int_{B_\varepsilon} (u_\varepsilon - u)u \,. \tag{3.43}$$

From the Cauchy-Schwarz inequality and Lemma 1.1, the remainder $\mathcal{E}_3(\varepsilon)$ can be bounded as follows:

$$\begin{aligned}
|\mathcal{E}_3(\varepsilon)| &\leq C_1\|u\|_{L^2(B_\varepsilon)}\|u_\varepsilon - u\|_{L^2(B_\varepsilon)} \\
&\leq C_2\varepsilon\|u_\varepsilon - u\|_{L^2(B_\varepsilon)} \\
&\leq C_3\varepsilon^{1+\delta}\|u_\varepsilon - u\|_{H^1(\Omega)} \\
&\leq C_4\varepsilon^{2+\delta} = o(\varepsilon^2),
\end{aligned} \tag{3.44}$$

with $0 < \delta < 1$, where we have also used Lemma 3.1 together with the interior elliptic regularity of function u. Therefore, *expansion* (3.42) can be rewritten as

$$\mathscr{J}_\varepsilon(u_\varepsilon) - \mathscr{J}(u) = \pi\varepsilon^2(1-\beta)|u(\widehat{x})|^2 + \sum_{i=3}^{4}\mathscr{E}_i(\varepsilon) . \qquad (3.45)$$

The remainder $\mathscr{E}_4(\varepsilon)$ is defined as

$$\mathscr{E}_4(\varepsilon) = (1-\beta)\int_{B_\varepsilon}(|u|^2 - |u(\widehat{x})|^2) , \qquad (3.46)$$

which can be trivially bounded as follows:

$$|\mathscr{E}_4(\varepsilon)| \leq C\varepsilon^3 = O(\varepsilon^3) , \qquad (3.47)$$

where we have used again the Cauchy-Schwarz inequality and the interior elliptic regularity of function u. According to the estimates (3.44) and (3.47), the remainders $\mathscr{E}_3(\varepsilon)$ and $\mathscr{E}_4(\varepsilon)$ are of order $o(\varepsilon^2)$. Therefore, from the expansion (3.45) we promptly identify the topological derivative as

$$\mathscr{T}_\beta(\widehat{x}) = (1-\beta)|u(\widehat{x})|^2 \quad \forall\,\widehat{x} \in \Omega , \qquad (3.48)$$

where we have chosen $f_\beta(\varepsilon) := f(\varepsilon) = \pi\varepsilon^2$.

3.4.3 Perturbation on the Higher Order Term

Finally, we set $\beta = \gamma = 1$ In this case, the variation of the compliance shape function (3.30) is given by

$$\mathscr{J}_\varepsilon(u_\varepsilon) - \mathscr{J}(u) = (1-\alpha)\int_{B_\varepsilon}\nabla u_\varepsilon \cdot \nabla u . \qquad (3.49)$$

By using the same strategy as before, we sum and subtract the term

$$(1-\alpha)\int_{B_\varepsilon}\|\nabla u\|^2 , \qquad (3.50)$$

allowing to rewrite (3.49) as follows:

$$\mathscr{J}_\varepsilon(u_\varepsilon) - \mathscr{J}(u) = (1-\alpha)\int_{B_\varepsilon}\|\nabla u\|^2 + \mathscr{I}(\varepsilon) . \qquad (3.51)$$

The term $\mathscr{I}(\varepsilon)$ is defined as

$$\mathscr{I}(\varepsilon) = (1 - \alpha) \int_{B_\varepsilon} \nabla(u_\varepsilon - u) \cdot \nabla u , \tag{3.52}$$

which can be bounded as follows:

$$|\mathscr{I}(\varepsilon)| \leq C_1 \|\nabla u\|_{L^2(B_\varepsilon)} \|\nabla(u_\varepsilon - u)\|_{L^2(B_\varepsilon)}$$
$$\leq C_2 \varepsilon \|u_\varepsilon - u\|_{H^1(\Omega)} \leq C_3 \varepsilon^2 = O(\varepsilon^2) , \tag{3.53}$$

where we have used Lemma 3.1, together with the interior elliptic regularity of function u. Since the above estimate cannot be improved, then there is a nontrivial term of order $O(\varepsilon^2)$ hidden in (3.52). On the other hand, according to Lemma 3.1, a leading term of order $O(\varepsilon^2)$ is expected. Therefore, in order to extract this term, we need to develop u_ε asymptotically with respect to the small parameter ε. The basic idea consists in postulating an *ansatz* for u_ε in the form [58]

$$u_\varepsilon(x) = u(x) + w_\varepsilon(x) + \tilde{u}_\varepsilon(x). \tag{3.54}$$

According to (3.5) and (3.6), $\beta_\varepsilon(x) = \gamma_\varepsilon(x) = 1$ for all $x \in \Omega$, since we have fixed $\beta = \gamma = 1$. Therefore, the state equation associated with the perturbed problem (3.9) can be written as

$$- \mathrm{div}\,(\alpha_\varepsilon \nabla u_\varepsilon) + u_\varepsilon = b. \tag{3.55}$$

After replacing the ansatz (3.54) in the new state equation (3.55), we obtain

$$b = -\mathrm{div}\,(\alpha_\varepsilon \nabla u_\varepsilon) + u_\varepsilon$$
$$= -\mathrm{div}\,(\alpha_\varepsilon \nabla u) + u - \mathrm{div}\,(\alpha_\varepsilon \nabla w_\varepsilon) + w_\varepsilon - \mathrm{div}\,(\alpha_\varepsilon \nabla \tilde{u}_\varepsilon) + \tilde{u}_\varepsilon . \tag{3.56}$$

With help of the contrast (3.4), the term $\mathrm{div}\,(\alpha_\varepsilon \nabla u)$ has to be interpreted as follows:

$$\mathrm{div}\,(\alpha_\varepsilon \nabla u) = \Delta u|_{\Omega \setminus \overline{B_\varepsilon}} + \alpha \Delta u|_{B_\varepsilon} \pm \Delta u|_{B_\varepsilon}$$
$$= \Delta u - (1 - \alpha)\Delta u|_{B_\varepsilon} , \tag{3.57}$$

since $\mathrm{div}\,(\nabla u) = \Delta u$. Now, we can replace this last result in (3.56) to obtain

$$b = -\Delta u + u + (1 - \alpha)\Delta u|_{B_\varepsilon} - \mathrm{div}\,(\alpha_\varepsilon \nabla w_\varepsilon) + w_\varepsilon - \mathrm{div}\,(\alpha_\varepsilon \nabla \tilde{u}_\varepsilon) + \tilde{u}_\varepsilon . \tag{3.58}$$

Therefore, from the state equation associated with the unperturbed problem (3.3) it follows that

$$- \mathrm{div}\,(\alpha_\varepsilon \nabla w_\varepsilon) + w_\varepsilon - \mathrm{div}\,(\alpha_\varepsilon \nabla \tilde{u}_\varepsilon) + \tilde{u}_\varepsilon + (1 - \alpha)\Delta u|_{B_\varepsilon} = 0. \tag{3.59}$$

Since we have some freedom to choose w_ε and $\tilde{u}_\varepsilon$, the following problems are defined:

$$- \operatorname{div}(\alpha_\varepsilon \nabla w_\varepsilon) = 0 \quad \text{and} \quad - \operatorname{div}(\alpha_\varepsilon \nabla \tilde{u}_\varepsilon) + \tilde{u}_\varepsilon = -(1-\alpha)\Delta u_{|_{B_\varepsilon}} - w_\varepsilon \ . \quad (3.60)$$

We have to complement the above problems with their associated boundary and transmission conditions. Thus, let us come back to the ansatz (3.54). After evaluating its gradient, we have

$$\nabla u_\varepsilon(x) = \nabla u(x) + \nabla w_\varepsilon(x) + \nabla \tilde{u}_\varepsilon(x) \ , \quad (3.61)$$

which can be rewritten as

$$\nabla u_\varepsilon(x) = \nabla u(\hat{x}) + (\nabla u(x) - \nabla u(\hat{x})) + \nabla w_\varepsilon(x) + \nabla \tilde{u}_\varepsilon(x) \ . \quad (3.62)$$

According to (3.9), the jump condition on the interface ∂B_ε is given by

$$\partial_n u_\varepsilon|_{\Omega \setminus \overline{B_\varepsilon}} - \alpha \partial_n u_\varepsilon|_{B_\varepsilon} = 0 \ . \quad (3.63)$$

Therefore, by combining the last two equations we have

$$(1-\alpha)\nabla u(\hat{x}) \cdot n + (1-\alpha)(\nabla u(x) - \nabla u(\hat{x})) \cdot n$$
$$+ \partial_n w_\varepsilon(x)|_{\Omega \setminus \overline{B_\varepsilon}} - \alpha \partial_n w_\varepsilon(x)|_{B_\varepsilon} + \partial_n \tilde{u}_\varepsilon(x)|_{\Omega \setminus \overline{B_\varepsilon}} - \alpha \partial_n \tilde{u}_\varepsilon(x)|_{B_\varepsilon} = 0 \ .$$
$$(3.64)$$

Thus, we can choose w_ε such that

$$\partial_n w_\varepsilon(x)|_{\Omega \setminus \overline{B_\varepsilon}} - \alpha \partial_n w_\varepsilon(x)|_{B_\varepsilon} = -(1-\alpha)\nabla u(\hat{x}) \cdot n \quad \text{on} \quad \partial B_\varepsilon \ . \quad (3.65)$$

From the above result together with (3.60)-left, the following *exterior problem* is considered, and formally obtained as $\varepsilon \to 0$: Find w_ε, such that

$$\begin{cases} \operatorname{div}(\alpha_\varepsilon \nabla w_\varepsilon) = 0 & \text{in} \ \mathbb{R}^2 \ , \\ w_\varepsilon \to 0 & \text{at} \ \infty \ , \\ \left. \begin{array}{r} w_\varepsilon|_{\mathbb{R}^2 \setminus \overline{B_\varepsilon}} - w_\varepsilon|_{B_\varepsilon} = 0 \\ \partial_n w_\varepsilon|_{\mathbb{R}^2 \setminus \overline{B_\varepsilon}} - \alpha \partial_n w_\varepsilon|_{B_\varepsilon} = \hat{v} \end{array} \right\} & \text{on} \ \partial B_\varepsilon \ , \end{cases} \quad (3.66)$$

with $\hat{v} = -(1-\alpha)\nabla u(\hat{x}) \cdot n$. The above boundary value problem admits an *explicit solution*, namely

$$w_\varepsilon(x)|_{\mathbb{R}^2 \setminus \overline{B_\varepsilon}} = \frac{1-\alpha}{1+\alpha} \frac{\varepsilon^2}{\|x - \hat{x}\|^2} \nabla u(\hat{x}) \cdot (x - \hat{x}) \ , \quad (3.67)$$

$$w_\varepsilon(x)|_{B_\varepsilon} = \frac{1-\alpha}{1+\alpha} \nabla u(\hat{x}) \cdot (x - \hat{x}) \ . \quad (3.68)$$

Now we can construct $\tilde{u}_\varepsilon$ in such a way that it compensates for the discrepancies introduced by the higher order terms in ε as well as by the boundary layer w_ε on the exterior boundary $\partial\Omega$. Therefore, from (3.60)-right, the *remainder* $\tilde{u}_\varepsilon$ must be the solution to the following boundary value problem: Find $\tilde{u}_\varepsilon$, such that

$$
\begin{cases}
-\mathrm{div}\,(\alpha_\varepsilon \nabla \tilde{u}_\varepsilon) + \tilde{u}_\varepsilon = f_\varepsilon & \text{in } \Omega, \\
\tilde{u}_\varepsilon = g_\varepsilon & \text{on } \partial\Omega, \\
\left.\tilde{u}_\varepsilon\right|_{\Omega \setminus \overline{B_\varepsilon}} - \left.\tilde{u}_\varepsilon\right|_{B_\varepsilon} = 0 \\
\left.\partial_n \tilde{u}_\varepsilon\right|_{\Omega \setminus \overline{B_\varepsilon}} - \alpha \partial_n \tilde{u}_\varepsilon|_{B_\varepsilon} = h_\varepsilon
\end{cases} \Biggr\} \text{ on } \partial B_\varepsilon,
\tag{3.69}
$$

where $f_\varepsilon = -(1-\alpha)\Delta u|_{B_\varepsilon} - w_\varepsilon$, $g_\varepsilon = -w_\varepsilon|_{\partial\Omega}$ and $h_\varepsilon = \tilde{v} \cdot n$, with the vector function $\tilde{v}(x) = -(1-\alpha)(\nabla u(x) - \nabla u(\hat{x}))$. Before proceeding, let us state the following result:

Lemma 3.2 *Let $\tilde{u}_\varepsilon$ be the solution to (3.69) or equivalently the solution to the following variational problem:*

$$
\tilde{u}_\varepsilon \in \mathscr{U}_\varepsilon : \int_\Omega \alpha_\varepsilon \nabla \tilde{u}_\varepsilon \cdot \nabla \eta + \int_\Omega \tilde{u}_\varepsilon \eta = \int_\Omega f_\varepsilon \eta + \int_{\partial B_\varepsilon} h_\varepsilon \eta \quad \forall \eta \in H_0^1(\Omega),
\tag{3.70}
$$

where the set $\mathscr{U}_\varepsilon$ is defined as

$$
\mathscr{U}_\varepsilon := \{\varphi \in H^1(\Omega) : \varphi|_{\partial\Omega} = g_\varepsilon\}.
\tag{3.71}
$$

Then, the following estimate for the remainder $\tilde{u}_\varepsilon$ holds true:

$$
\|\tilde{u}_\varepsilon\|_{H^1(\Omega)} \le C\varepsilon^2 \sqrt{|\log \varepsilon|},
\tag{3.72}
$$

with constant C independent of the small parameter ε.

Proof From the definition of the source term $f_\varepsilon = -(1-\alpha)\Delta u|_{B_\varepsilon} - w_\varepsilon$, there is

$$
\int_\Omega f_\varepsilon \eta = -(1-\alpha)\int_{B_\varepsilon} \Delta u\, \eta - \int_\Omega w_\varepsilon \eta.
\tag{3.73}
$$

In addition, since function $h_\varepsilon = \tilde{v} \cdot n$, with n used to denote the unit normal vector field on ∂B_ε pointing toward to the center of the inclusion, it follows that

$$
\int_{\partial B_\varepsilon} h_\varepsilon \eta = \int_{\partial B_\varepsilon} \tilde{v} \cdot n\, \eta = -\int_{B_\varepsilon} \mathrm{div}\,(\tilde{v}\eta) = -\int_{B_\varepsilon} \mathrm{div}\,(\tilde{v})\eta - \int_{B_\varepsilon} \tilde{v} \cdot \nabla \eta
$$

$$
= (1-\alpha)\int_{B_\varepsilon} \Delta u\, \eta + (1-\alpha)\int_{B_\varepsilon} (\nabla u - \nabla u(\hat{x})) \cdot \nabla \eta,
\tag{3.74}
$$

where we have taken into account that $\tilde{v} = -(1-\alpha)(\nabla u - \nabla u(\hat{x}))$. From these last two results, the variational form (3.70) can be rewritten as follows:

$$\widetilde{u}_\varepsilon \in \mathscr{U}_\varepsilon \ : \ \int_\Omega \alpha_\varepsilon \nabla \widetilde{u}_\varepsilon \cdot \nabla \eta + \int_\Omega \widetilde{u}_\varepsilon \eta = -\int_{B_\varepsilon} \widetilde{v} \cdot \nabla \eta - \int_\Omega w_\varepsilon \eta \quad \forall \eta \in H_0^1(\Omega) \ .$$

(3.75)

Now, let us take $\eta = \widetilde{u}_\varepsilon - \varphi_\varepsilon$ as test function in (3.75), where $\varphi_\varepsilon \in \mathscr{U}_\varepsilon$ is the lifting of the Dirichlet boundary data g_ε on $\partial\Omega$. Therefore

$$\int_\Omega \alpha_\varepsilon \|\nabla \widetilde{u}_\varepsilon\|^2 + \int_\Omega |\widetilde{u}_\varepsilon|^2 = \int_{\partial\Omega} g_\varepsilon \partial_n \widetilde{u}_\varepsilon - \int_{B_\varepsilon} \widetilde{v} \cdot \nabla \widetilde{u}_\varepsilon - \int_\Omega w_\varepsilon \widetilde{u}_\varepsilon \ .$$

(3.76)

From the *Cauchy-Schwarz inequality* and the *trace theorem*, there is

$$\left| \int_{\partial\Omega} g_\varepsilon \partial_n \widetilde{u}_\varepsilon \right| \leq \|g_\varepsilon\|_{H^{1/2}(\partial\Omega)} \|\partial_n \widetilde{u}_\varepsilon\|_{H^{-1/2}(\partial\Omega)}$$

$$\leq \|g_\varepsilon\|_{H^{1/2}(\partial\Omega)} \|\nabla \widetilde{u}_\varepsilon\|_{L^2(\Omega)} \ .$$

(3.77)

Since $g_\varepsilon = -w_{\varepsilon|\partial\Omega}$, then, according to the explicit solution (3.67), g_ε has order $O(\varepsilon^2)$ on $\partial\Omega$. Therefore

$$\left| \int_{\partial\Omega} g_\varepsilon \partial_n \widetilde{u}_\varepsilon \right| \leq C_1 \varepsilon^2 \|\widetilde{u}_\varepsilon\|_{H^1(\Omega)}.$$

(3.78)

By taking into account the definition $\widetilde{v}(x) = -(1-\alpha)(\nabla u(x) - \nabla u(\widehat{x}))$, there is

$$\left| \int_{B_\varepsilon} \widetilde{v} \cdot \nabla \widetilde{u}_\varepsilon \right| \leq \|\widetilde{v}\|_{L^2(B_\varepsilon)} \|\nabla \widetilde{u}_\varepsilon\|_{L^2(B_\varepsilon)}$$

$$\leq C_2 \|\nabla u(x) - \nabla u(\widehat{x})\|_{L^2(B_\varepsilon)} \|\nabla \widetilde{u}_\varepsilon\|_{L^2(B_\varepsilon)}$$

$$\leq C_3 \|x - \widehat{x}\|_{L^2(B_\varepsilon)} \|\nabla \widetilde{u}_\varepsilon\|_{L^2(B_\varepsilon)} \leq C_4 \varepsilon^2 \|\widetilde{u}_\varepsilon\|_{H^1(\Omega)} \ , \quad (3.79)$$

where we have used the Cauchy-Schwarz inequality and the interior elliptic regularity of function u. Let us consider again the Cauchy-Schwarz inequality to obtain

$$\left| \int_\Omega w_\varepsilon \widetilde{u}_\varepsilon \right| \leq C_5 \|w_\varepsilon\|_{L^2(\Omega)} \|\widetilde{u}_\varepsilon\|_{L^2(\Omega)}$$

$$\leq C_6 \varepsilon^2 \sqrt{|\log \varepsilon|} \|\widetilde{u}_\varepsilon\|_{H^1(\Omega)},$$

(3.80)

where we have used the explicit solution for w_ε given by (3.67) and (3.68). From these results it follows that

$$\int_\Omega \alpha_\varepsilon \|\nabla \widetilde{u}_\varepsilon\|^2 + \int_\Omega |\widetilde{u}_\varepsilon|^2 \leq C_7 \varepsilon^2 \sqrt{|\log \varepsilon|} \|\widetilde{u}_\varepsilon\|_{H^1(\Omega)}.$$

(3.81)

Finally, from the *coercivity* of the bilinear form on the left-hand side of the above inequality, namely

$$c\|\tilde{u}_\varepsilon\|^2_{H^1(\Omega)} \leq \int_\Omega \alpha_\varepsilon \|\nabla \tilde{u}_\varepsilon\|^2 + \int_\Omega |\tilde{u}_\varepsilon|^2 , \tag{3.82}$$

we obtain the result, with constant $C = C_7/c$. $\square$

Now, we can develop the integral $\mathscr{I}(\varepsilon)$ given by (3.52) in power of ε. In fact, by replacing the ansatz (3.54) into (3.52) we have

$$\mathscr{I}(\varepsilon) = (1-\alpha) \int_{B_\varepsilon} \nabla(w_\varepsilon + \tilde{u}_\varepsilon) \cdot \nabla u$$

$$= (1-\alpha) \int_{B_\varepsilon} \nabla w_\varepsilon \cdot \nabla u + \mathscr{E}_5(\varepsilon) . \tag{3.83}$$

The remainder $\mathscr{E}_5(\varepsilon)$ is defined as

$$\mathscr{E}_5(\varepsilon) = (1-\alpha) \int_{B_\varepsilon} \nabla \tilde{u}_\varepsilon \cdot \nabla u . \tag{3.84}$$

From the Cauchy-Schwarz inequality there is

$$|\mathscr{E}_5(\varepsilon)| \leq C_1 \|\nabla u\|_{L^2(B_\varepsilon)} \|\nabla \tilde{u}_\varepsilon\|_{L^2(B_\varepsilon)} \leq C_2 \varepsilon \|\nabla \tilde{u}_\varepsilon\|_{L^2(\Omega)} , \tag{3.85}$$

where we have used the interior elliptic regularity of function u. By taking into account Lemma 3.2, there is

$$|\mathscr{E}_5(\varepsilon)| \leq C_3 \varepsilon \|\tilde{u}_\varepsilon\|_{H^1(\Omega)} \leq C_4 \varepsilon^3 \sqrt{|\log \varepsilon|} = o(\varepsilon^2) . \tag{3.86}$$

Now, we can come back to the *expansion* (3.51), which can be written as

$$\mathscr{J}_\varepsilon(u_\varepsilon) - \mathscr{J}(u) = (1-\alpha) \int_{B_\varepsilon} \|\nabla u\|^2 + (1-\alpha) \int_{B_\varepsilon} \nabla w_\varepsilon \cdot \nabla u + \mathscr{E}_5(\varepsilon)$$

$$= (1-\alpha) \int_{B_\varepsilon} \|\nabla u(\hat{x})\|^2 + (1-\alpha) \int_{B_\varepsilon} \nabla w_\varepsilon \cdot \nabla u(\hat{x}) + \sum_{i=5}^{7} \mathscr{E}_i(\varepsilon)$$

$$= \pi \varepsilon^2 (1-\alpha) \|\nabla u(\hat{x})\|^2 + \pi \varepsilon^2 \frac{(1-\alpha)^2}{1+\alpha} \|\nabla u(\hat{x})\|^2 + \sum_{i=5}^{7} \mathscr{E}_i(\varepsilon)$$

$$= 2\pi \varepsilon^2 \frac{1-\alpha}{1+\alpha} \|\nabla u(\hat{x})\|^2 + \sum_{i=5}^{7} \mathscr{E}_i(\varepsilon) , \tag{3.87}$$

where we have used the explicit solution (3.68), namely

$$\nabla w_\varepsilon(x)_{|B_\varepsilon} = \frac{1-\alpha}{1+\alpha} \nabla u(\widehat{x}) \,. \tag{3.88}$$

The remainders $\mathscr{E}_6(\varepsilon)$ and $\mathscr{E}_7(\varepsilon)$ are respectively defined as

$$\mathscr{E}_6(\varepsilon) = (1-\alpha) \int_{B_\varepsilon} \left(\|\nabla u\|^2 - \|\nabla u(\widehat{x})\|^2 \right) \,, \tag{3.89}$$

$$\mathscr{E}_7(\varepsilon) = (1-\alpha) \int_{B_\varepsilon} \nabla w_\varepsilon \cdot (\nabla u - \nabla u(\widehat{x})) \,, \tag{3.90}$$

which can be trivially bounded as follows:

$$|\mathscr{E}_6(\varepsilon)| \le C_1 \varepsilon^3 = O\left(\varepsilon^3 \right) \,, \tag{3.91}$$

$$|\mathscr{E}_7(\varepsilon)| \le C_2 \varepsilon^3 = O\left(\varepsilon^3 \right) \,, \tag{3.92}$$

where we have used the interior elliptic regularity of function u and the explicit solution (3.88). According to the estimates (3.86) and (3.91), the remainders $\mathscr{E}_5(\varepsilon)$, $\mathscr{E}_6(\varepsilon)$, and $\mathscr{E}_7(\varepsilon)$ are of order $o(\varepsilon^2)$. Therefore, from the expansion (3.87) we promptly identify the topological derivative as

$$\mathscr{T}_\alpha(\widehat{x}) = 2 \frac{1-\alpha}{1+\alpha} \|\nabla u(\widehat{x})\|^2 \quad \forall \widehat{x} \in \Omega \,, \tag{3.93}$$

where we have chosen $f_\alpha(\varepsilon) := f(\varepsilon) = \pi \varepsilon^2$.

Remark 3.1 (Pólya-Szegö Polarization Tensor) In the case of arbitrary shaped inclusions $\omega_\varepsilon(\widehat{x}) = \widehat{x} + \varepsilon \omega$, the exterior problem (3.66), written in its variational form, reads:

$$w_\varepsilon \in \mathscr{W} \,: \int_{\mathbb{R}^2} \alpha_\varepsilon \nabla w_\varepsilon \cdot \nabla \eta = (1-\alpha) \nabla u(\widehat{x}) \cdot \int_{\omega_\varepsilon} \nabla \eta \quad \forall \eta \in \mathscr{W} \,, \tag{3.94}$$

where the quotient space $\mathscr{W}$ is defined as $\mathscr{W} := \{\varphi \in H^1(\mathbb{R}^2)/\mathbb{R}\}$. Now, let us introduce the notation $w(\varepsilon^{-1} x) := \varepsilon^{-1} w_\varepsilon(x)$ and the change of variable $\xi = \varepsilon^{-1} x$. From the linearity of the above variational problem, $w(\xi)$ can be written in terms of the components of $\nabla u(\widehat{x})$ as follows: $w(\xi) = (\nabla u(\widehat{x}))_i v^{(i)}(\xi)$, where $(\nabla u(\widehat{x}))_i = \nabla u(\widehat{x}) \cdot e_i$, with e_i used to denote the canonical basis of $\mathbb{R}^2$. By taking into account these elements into the above exterior problem, the following set of canonical variational problems can be introduced:

$$v^{(i)} \in \mathscr{W} \,: \int_{\mathbb{R}^2} \alpha_\omega \nabla_\xi v^{(i)} \cdot \nabla_\xi \eta = (1-\alpha) e_i \cdot \int_\omega \nabla_\xi \eta \quad \forall \eta \in \mathscr{W} \,, \tag{3.95}$$

where the contrast α_ω is defined as $\alpha_\omega = 1$ in $\mathbb{R}^2 \setminus \omega$ and $\alpha_\omega = \alpha$ in ω. On the other hand, the variation of the compliance shape function given by (3.87) can be rewritten as

$$\mathscr{J}_\varepsilon(u_\varepsilon) - \mathscr{J}(u) = (1-\alpha)\int_{\omega_\varepsilon} (\nabla u(\widehat{x}) + \nabla w_\varepsilon) \cdot \nabla u(\widehat{x}) + o\left(\varepsilon^2\right)$$

$$= (1-\alpha)\varepsilon^2 \int_\omega (\nabla u(\widehat{x}) + \nabla_\xi w) \cdot \nabla u(\widehat{x}) + o\left(\varepsilon^2\right), \quad (3.96)$$

since $w_\varepsilon(x) = \varepsilon w(\varepsilon^{-1}x)$ and $\nabla_\xi w(\xi) = \varepsilon \nabla w(\varepsilon^{-1}x)$. In addition, the gradient of $w(\xi)$ with respect to ξ can be written as

$$\nabla_\xi w = (\nabla u(\widehat{x}) \cdot e_j)\nabla_\xi v^{(j)}$$

$$= (\nabla u(\widehat{x}) \cdot e_j)(\nabla_\xi v^{(j)})_i \, e_i$$

$$= (\nabla_\xi v^{(j)})_i (e_i \otimes e_j)\nabla u(\widehat{x}) . \quad (3.97)$$

After replacing this decomposition into the above expansion we obtain

$$\mathscr{J}_\varepsilon(u_\varepsilon) - \mathscr{J}(u) = -|\omega|\varepsilon^2 \mathrm{P}_\alpha \nabla u(\widehat{x}) \cdot \nabla u(\widehat{x}) + o(\varepsilon^2) . \quad (3.98)$$

The Pólya-Szegö *Polarization Tensor* is defined as [5]

$$\mathrm{P}_\alpha := -(1-\alpha)\left(\mathrm{I} + \frac{1}{|\omega|}\int_\omega (\nabla_\xi v^{(j)})_i (e_i \otimes e_j)\right), \quad (3.99)$$

where I is used to denote the second order identity tensor. The set $\omega \subset \mathbb{R}^2$ is a reference domain representing an inclusion of arbitrary shape and $|\omega|$ is the Lebesgue measure of ω. Finally, functions $v^{(j)}$ are solutions to the set of canonical variational problems (3.95) written in terms of the fast variable $\xi \in \mathbb{R}^2$, which have to be solved for each canonical direction $e_1 = (1, 0)^\top$ and $e_2 = (0, 1)^\top$. Note that the set of canonical variational problems (3.95) and the integral (3.99) can be evaluated numerically. Therefore, this approach induces a numerical procedure for evaluating the polarization tensor associated with arbitrary shaped inclusions embedded into an anisotropic and heterogeneous medium.

3.5   Summary of the Results

In this chapter the topological derivative of the compliance shape functional associated with a modified Helmholtz problem, with respect to the nucleation of a small inclusion endowed with different material properties from the background, has been derived. We have considered three different scenarios, which are: perturbation on the right-hand side, perturbation on the lower order term, and perturbation on the higher order term. After collecting the results derived in Sect. 3.4, we have

$$\psi(\chi_\varepsilon(\widehat{x})) = \psi(\chi) + f_\alpha(\varepsilon)\mathscr{T}_\alpha(\widehat{x}) + f_\beta(\varepsilon)\mathscr{T}_\beta(\widehat{x}) + f_\gamma(\varepsilon)\mathscr{T}_\gamma(\widehat{x}) + \mathscr{E}(\varepsilon), \quad (3.100)$$

remembering that $\psi(\chi) = \mathscr{J}(u)$ and $\psi(\chi_\varepsilon) = \mathscr{J}_\varepsilon(u_\varepsilon)$ are defined by (3.1) and (3.7), respectively. In addition, functions $f_\alpha(\varepsilon) = f_\beta(\varepsilon) = f_\gamma(\varepsilon) = \pi\varepsilon^2$ and the remainder $\mathscr{E}(\varepsilon)$ is defined as

$$\mathscr{E}(\varepsilon) := \sum_{i=1}^{7} \mathscr{E}_i(\varepsilon) = o(\varepsilon^2), \quad (3.101)$$

since each $\mathscr{E}_i(\varepsilon) = o(\varepsilon^2)$, for $i = 1, \cdots, 7$. Finally, the topological derivatives $\mathscr{T}_\alpha(\widehat{x})$, $\mathscr{T}_\beta(\widehat{x})$ and $\mathscr{T}_\gamma(\widehat{x})$ are given by (3.39), (3.48), and (3.93), respectively. These sensitivities can be interpreted as partial topological derivatives with respect to each contrast α, β, and γ. Therefore, from the basic rules of differential calculus, expansion (3.100) can be written as

$$\psi(\chi_\varepsilon(\widehat{x})) = \psi(\chi) + \pi\varepsilon^2 \mathscr{T}(\widehat{x}) + o(\varepsilon^2), \quad (3.102)$$

where $\mathscr{T}(\widehat{x})$ is given by the sum

$$\mathscr{T}(\widehat{x}) = \mathscr{T}_\alpha(\widehat{x}) + \mathscr{T}_\beta(\widehat{x}) + \mathscr{T}_\gamma(\widehat{x})$$
$$= 2\frac{1-\alpha}{1+\alpha}\|\nabla u(\widehat{x})\|^2 + (1-\beta)|u(\widehat{x})|^2 - 2(1-\gamma)b(\widehat{x})u(\widehat{x}), \quad (3.103)$$

which represents the total *topological derivative* of the shape functional ψ.

3.6 Exercises

1. From the weak formulation (3.2) derive the strong form (3.3).
2. From the weak formulation (3.8) derive the strong form (3.9) and discuss the transmission condition on the interface ∂B_ε.
3. By using separation of variable technique, find the explicit solution for the exterior problem (3.66).
4. Show that the solution w_ε of the exterior problem (3.66) is bounded as follows: $\|w_\varepsilon\|_{L^2(\Omega)} \le C\varepsilon^2\sqrt{|\log\varepsilon|}$, which has been used to derive estimation (3.80).
   Hint: Define a big ball $B_R(\widehat{x})$ of radius R and center at $\widehat{x} \in \Omega$, such that $\overline{\Omega} \subset B_R(\widehat{x})$. Then, it follows immediately that

$$\|w_\varepsilon\|_{L^2(\Omega)} \le \|w_\varepsilon\|_{L^2(B_R)} = \left(\|w_\varepsilon\|_{L^2(B_R\setminus\overline{B_\varepsilon})}^2 + \|w_\varepsilon\|_{L^2(B_\varepsilon)}^2\right)^{1/2}.$$

In a polar coordinate system (r, θ) with center at $\widehat{x}$, the explicit solution w_ε from (3.67) and (3.68) can be written, respectively, as

- for $\varepsilon \leq r < R$

$$w^e(r, \theta) = \frac{1-\alpha}{1+\alpha}\varepsilon^2 r^{-1}(\varphi_1 \cos(\theta) + \varphi_2 \sin(\theta));$$

- for $0 < r < \varepsilon$

$$w^i(r, \theta) = \frac{1-\alpha}{1+\alpha} r \left(\varphi_1 \cos(\theta) + \varphi_2 \sin(\theta)\right),$$

where $(\varphi_1, \varphi_2)^\top := \nabla u(\widehat{x}) \in \mathbb{R}^2$.

Then, compute by hand the integrals

$$\|w_\varepsilon\|^2_{L^2(B_R \setminus \overline{B_\varepsilon})} = \int_0^{2\pi} \left(\int_\varepsilon^R |w^e(r, \theta)|^2 r dr \right) d\theta,$$

$$\|w_\varepsilon\|^2_{L^2(B_\varepsilon)} = \int_0^{2\pi} \left(\int_0^\varepsilon |w^i(r, \theta)|^2 r dr \right) d\theta.$$

5. From the strong formulation (3.69) derive the weak form (3.70). Then, derive the equality (3.76) by taking $\eta = \widetilde{u}_\varepsilon - \varphi_\varepsilon$ as test function in (3.75), with $\varphi_\varepsilon \in \mathscr{U}_\varepsilon$, where $\mathscr{U}_\varepsilon$ is defined by (3.71).
6. By taking $\omega_\varepsilon(\widehat{x}) = B_\varepsilon(\widehat{x})$ in Remark 3.1, show that the polarization tensor defined through (3.99) degenerates itself to $P_\alpha = -2\frac{1-\alpha}{1+\alpha}I$, which corroborates with the final result given by (3.93), provided that $|\omega| = \pi$ for circular inclusions.

Chapter 4
Domain Truncation Method

In this chapter, the domain decomposition technique is combined with the Steklov–Poincaré pseudo-differential boundary operator for asymptotic analysis purposes with respect to the small parameter associated with the size of the topological perturbation. As a fundamental result, the expansion of the energy coincides with the expansion of the Steklov–Poincaré operator on the boundary of the truncated domain, leading to the associated topological derivative. The proposed method is general and can be applied in the topological asymptotic analysis of a wide range of multi-physics and nonlinear problems. Actually, the method has been developed in the context of unilateral contact problems [89], which are governed by nonlinear variational inequalities. See the book by Novotny and Sokołowski [75], for instance. In addition, through this approach, the estimation of the remainders left by the topological asymptotic expansion are obtained in a subdomain free of singularities, which allows to construct the mathematical arguments needed to justify all derivations by using elementary tools from the analysis.

4.1 Coupled System

The geometrical domain $\mathscr{D} \subset \mathbb{R}^2$, with boundary $\Gamma = \partial \mathscr{D}$, is decomposed into two subdomains $\omega \subset \mathscr{D}$ and $\Omega = \mathscr{D} \setminus \overline{\omega}$ endowed with different physical properties, as shown in Fig. 4.1. The interaction between ω and Ω is governed by a transmission condition acting on the interface $\partial \omega$. Since the resulting model is complicated, the *domain decomposition technique* is used in the topological asymptotic analysis. The basic idea consists in introducing a fictitious ring $C_R(\varepsilon) := \{\varepsilon < \|x\| < R\}$ for asymptotic analysis purposes with respect to the small parameter $\varepsilon \to 0$. The associated expansion is written on the boundary ∂B_R of the ball $B_R = \{\|x\| < R\}$. As a fundamental result, the expansion of the energy in $C_R(\varepsilon)$ coincides with the expansion of the *Steklov–Poincaré* pseudo-differential operator

© The Author(s), under exclusive license to Springer Nature Switzerland AG 2020
A. A. Novotny, J. Sokołowski, *An Introduction to the Topological
Derivative Method*, SpringerBriefs in Mathematics,
https://doi.org/10.1007/978-3-030-36915-6_4

Fig. 4.1 Geometrical domain
$\mathscr{D}$ decomposed into $\omega \subset \mathscr{D}$
and $\Omega = \mathscr{D} \setminus \overline{\omega}$

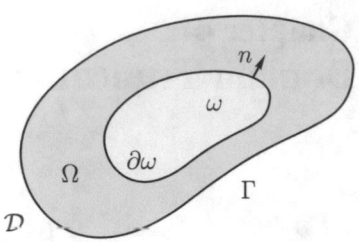

defined on the fictitious boundary ∂B_R, allowing for identifying the associated
topological derivative. Therefore, the influence of the singularity arising from the
limit passage $\varepsilon \to 0$ can be analyzed in the truncated domain free of singularities
through nonlocal boundary conditions on ∂B_R. In addition, the compactness results
used to justify the proposed topological asymptotic expansion are derived by using
classical *Fourier analysis* in the singularly perturbed geometrical domain $C_R(\varepsilon)$.
This approach simplifies the topological sensitivity analysis of the associated shape
functional.

4.1.1   Non-perturbed Problem

The *shape functional* defined in the non-perturbed geometrical domain Ω is given
by

$$\psi(\chi) := \mathscr{J}_\Omega(u) = \frac{1}{2} \int_\Omega |u - z_d|^2 \,, \tag{4.1}$$

where z_d is the target function and u is the solution of the following variational
problem: Find $u \in H_0^1(\mathscr{D})$, such that

$$\int_\omega \nabla u \cdot \nabla \eta + \int_\Omega \nabla u \cdot \nabla \eta + \int_\Omega u\eta = \int_\Omega b\eta \quad \forall \eta \in H_0^1(\mathscr{D})\,. \tag{4.2}$$

In the above equation, b represents a source term assumed to be smooth enough.
The strong formulation associated with the variational problem (4.2) is given by the
following coupled boundary value problem: Find w, such that

$$\begin{cases} -\Delta w = 0 \text{ in } \omega \,, \\ \quad\; w = \varphi \text{ on } \partial \omega \,, \end{cases} \tag{4.3}$$

where $\mathscr{S}(\varphi)$ on $\partial \omega$ is the *Steklov–Poincaré operator* associated with the non-
perturbed problem, which is defined as $\mathscr{S} : \varphi \in H^{1/2}(\partial \omega) \mapsto \partial_n u \in H^{-1/2}(\partial \omega)$.
In addition, the restriction of the solution to (4.2) over Ω is given by the following
boundary value problem: Find u, such that

$$\begin{cases} -\Delta u + u = b & \text{in } \Omega , \\ u = 0 & \text{on } \Gamma , \\ \partial_n u = \mathscr{S}(u) \text{ on } \partial\omega . \end{cases} \tag{4.4}$$

In order to simplify further analysis, let us introduce the *adjoint state* v, which is the solution of the following variational problem (see Sect. 1.2.1 for details): Find $v \in H_0^1(\mathscr{D})$, such that

$$\int_\omega \nabla v \cdot \nabla \eta + \int_\Omega \nabla v \cdot \nabla \eta + \int_\Omega v\eta = -\int_\Omega (u - z_d)\eta \quad \forall \eta \in H_0^1(\mathscr{D}) . \tag{4.5}$$

4.1.2 Perturbed Problem

We consider the same problem but now defined in the topologically perturbed domain $\mathscr{D}_\varepsilon(\widehat{x}) = \mathscr{D} \setminus \overline{B_\varepsilon(\widehat{x})}$, with $\widehat{x} \in \omega \subset \mathscr{D}$. Therefore, $\omega_\varepsilon(\widehat{x}) = \omega \setminus \overline{B_\varepsilon(\widehat{x})}$ represents the subdomain effectively perturbed. In this case, the shape functional is defined in a fix domain, namely

$$\psi(\chi_\varepsilon) := \mathscr{J}_\Omega(u_\varepsilon) = \frac{1}{2}\int_\Omega |u_\varepsilon - z_d|^2 , \tag{4.6}$$

where the function u_ε is the solution of the following variational problem: Find $u_\varepsilon \in \mathscr{V}_\varepsilon$, such that

$$\int_{\omega_\varepsilon} \nabla u_\varepsilon \cdot \nabla \eta + \int_\Omega \nabla u_\varepsilon \cdot \nabla \eta + \int_\Omega u_\varepsilon \eta = \int_\Omega b\eta \quad \forall \eta \in \mathscr{V}_\varepsilon . \tag{4.7}$$

The space $\mathscr{V}_\varepsilon$ is defined as

$$\mathscr{V}_\varepsilon := \{\varphi \in H^1(\mathscr{D}_\varepsilon) : \varphi_{|\partial\mathscr{D}} = 0\} . \tag{4.8}$$

The *strong form* associated with the variational problem (4.7) can be written in the form of the following coupled boundary value problem: Find w_ε, such that

$$\begin{cases} -\Delta w_\varepsilon = 0 \text{ in } \omega_\varepsilon , \\ w_\varepsilon = \varphi \text{ on } \partial\omega , \\ \partial_n w_\varepsilon = 0 \text{ on } \partial B_\varepsilon , \end{cases} \tag{4.9}$$

where $\mathscr{S}_\varepsilon(\varphi)$ on $\partial\omega$ is the *Steklov–Poincaré operator* associated with the perturbed problem, which is defined as $\mathscr{S}_\varepsilon : \varphi \in H^{1/2}(\partial\omega) \mapsto \partial_n u_\varepsilon \in H^{-1/2}(\partial\omega)$. Again, the restriction of the solution to (4.7) over Ω is given by the following boundary value problem: Find u_ε, such that

$$\begin{cases} -\Delta u_\varepsilon + u_\varepsilon = b & \text{in } \Omega \,, \\ \quad u_\varepsilon = 0 & \text{on } \Gamma \,, \\ \quad \partial_n u_\varepsilon = \mathscr{S}_\varepsilon(u_\varepsilon) & \text{on } \partial\omega \,. \end{cases} \qquad (4.10)$$

4.2   Domain Decomposition Technique

The Steklov–Poincaré operator is now applied for decomposing the singularly perturbed geometrical domain ω_ε into two subdomains. One of them is given by the truncated domain $\omega \setminus \overline{B_R}$, where $B_R = B_R(\widehat{x})$. The other one is precisely the ring $C_R(\varepsilon) := B_R \setminus \overline{B_\varepsilon}$ which contains the singularity, with $B_\varepsilon = B_\varepsilon(\widehat{x})$. See Fig. 4.2. Therefore, the Steklov–Poincaré operator defined on the fictitious boundary ∂B_R becomes dependent on the control parameter ε, namely

$$\mathscr{A}_\varepsilon : H^{1/2}(\partial B_R) \mapsto H^{-1/2}(\partial B_R) \,. \qquad (4.11)$$

The region $C_R(\varepsilon)$, which includes the singular domain perturbation B_ε, is selected for asymptotic analysis with respect to the small parameter $\varepsilon \to 0$ that governs the singularity. The result obtained from the asymptotic analysis in $C_R(\varepsilon)$ is evaluated on the fictitious boundary ∂B_R where the Steklov–Poincaré pseudo-differential operator is defined. Therefore, the asymptotic expansion is derived in the simplified domain represented by the ring $C_R(\varepsilon)$ enjoying radial symmetry, which is separated from the topological sensitivity analysis of the shape functional obtained in the truncated domain $\mathscr{D}_R = \mathscr{D} \setminus \overline{B_R}$ endowed with nonlocal boundary conditions governed by the Steklov–Poincaré operator on ∂B_R, but free of the singularities produced by B_ε with $\varepsilon \to 0$. Thus, the following variational problem is considered in the truncated domain $\mathscr{D}_R$: Find $u_\varepsilon \in \mathscr{V}_R$, such that

$$\int_{\omega \setminus \overline{B_R}} \nabla u_\varepsilon \cdot \nabla \eta + \int_{\partial B_R} \mathscr{A}_\varepsilon(u_\varepsilon)\eta + \int_\Omega \nabla u_\varepsilon \cdot \nabla \eta + \int_\Omega u_\varepsilon \eta = \int_\Omega b\eta \quad \forall \eta \in \mathscr{V}_R \,, \qquad (4.12)$$

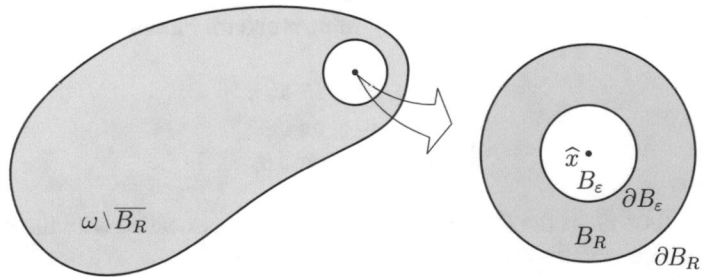

Fig. 4.2 Topologically perturbed geometrical domain ω_ε decomposed into $\omega \setminus \overline{B_R}$ and $C_R(\varepsilon) := B_R \setminus \overline{B_\varepsilon}$

whose nonlocal condition on the fictitious boundary ∂B_R is governed by the Steklov–Poincaré operator. The space $\mathcal{V}_R$ is defined as

$$\mathcal{V}_R := \{\varphi \in H^1(\mathcal{D}_R) : \varphi|_{\partial \mathcal{D}} = 0\} . \tag{4.13}$$

On the other hand, the singularity regarding the small parameter $\varepsilon \to 0$ is absorbed by the following boundary value problem defined in the ring $C_R(\varepsilon) = B_R \setminus \overline{B_\varepsilon}$: Find w_ε, such that

$$\begin{cases} -\Delta w_\varepsilon = 0 \text{ in } C_R(\varepsilon) , \\ w_\varepsilon = \varphi \text{ on } \partial B_R , \\ \partial_n w_\varepsilon = 0 \text{ on } \partial B_\varepsilon , \end{cases} \tag{4.14}$$

where $\mathscr{A}_\varepsilon(\varphi)$ on ∂B_R is the Steklov–Poincaré operator associated with the topologically perturbed problem, namely $\mathscr{A}_\varepsilon : \varphi \in H^{1/2}(\partial B_R) \mapsto \partial_n u_\varepsilon \in H^{-1/2}(\partial B_R)$.

Remark 4.1 The mapping $\varphi \mapsto \mathscr{A}_\varepsilon(\varphi)$ is defined as the Steklov–Poincaré operator associated with (4.14), so that the solution to the problem (4.12) is given by the restriction of the solution to (4.7) over the truncated domain $\mathcal{D}_R = \mathcal{D} \setminus \overline{B_R}$.

The energy regarding problem (4.14) enjoys the following property:

$$0 = -\int_{B_R \setminus \overline{B_\varepsilon}} \Delta w_\varepsilon \, w_\varepsilon = \int_{B_R \setminus \overline{B_\varepsilon}} \|\nabla w_\varepsilon\|^2 - \int_{\partial B_R} \partial_n w_\varepsilon \, w_\varepsilon$$

$$= \int_{B_R \setminus \overline{B_\varepsilon}} \|\nabla w_\varepsilon\|^2 - \int_{\partial B_R} \mathscr{A}_\varepsilon(\varphi) \, \varphi , \tag{4.15}$$

that is, the energy in the ring $C_R(\varepsilon)$ is equal to the energy associated with the Steklov–Poincaré operator on the fictitious boundary ∂B_R, namely

$$\int_{C_R(\varepsilon)} \|\nabla w_\varepsilon\|^2 = \int_{\partial B_R} \mathscr{A}_\varepsilon(\varphi) \, \varphi . \tag{4.16}$$

In addition, since the operator $\mathscr{A}_\varepsilon$ is symmetric, we have

$$\int_{C_R(\varepsilon)} \|\nabla w_\varepsilon\|^2 = \langle \mathscr{A}_\varepsilon(\varphi), \varphi \rangle_{(H^{-1/2} \times H^{1/2})(\partial B_R)} := \langle \mathscr{A}_\varepsilon(\varphi), \varphi \rangle_{\partial B_R} . \tag{4.17}$$

Thus, the asymptotic expansion of the Steklov–Poincaré operator coincides with the asymptotic expansion of the energy in the ring $C_R(\varepsilon)$. In particular, the energy in the ring $C_R(\varepsilon)$ admits an asymptotic expansion with respect to the small parameter $\varepsilon \to 0$ of the form:

$$\int_{C_R(\varepsilon)} \|\nabla w_\varepsilon\|^2 = \int_{B_R} \|\nabla w\|^2 - 2\pi \varepsilon^2 \|\nabla w(\widehat{x})\|^2 + o(\varepsilon^2) . \tag{4.18}$$

See Chap. 2, expansion (2.37) for $b = 0$ in the neighborhood of $\widehat{x} \in \Omega$.

In general, the density energy functional $H^1(B_R) \ni \varphi \mapsto \|\nabla\varphi(\widehat{x})\|^2 \in \mathbb{R}$ is not continuous at the point $\widehat{x}$. Therefore, the energy density is replaced by the continuous bilinear form $H^1(B_R) \ni \varphi \mapsto \langle \mathscr{B}(\varphi), \varphi \rangle_{\partial B_R} \in \mathbb{R}$. For the Laplacian into two spatial dimensions, the solution of the unperturbed problem w is harmonic in the neighborhood of $\widehat{x}$. Thus, a continuous bilinear form with respect to the norm $H^1(B_R)$ can be defined such that the following equality holds true:

$$\langle \mathscr{B}(w), w \rangle_{\partial B_R} = 2\|\nabla w(\widehat{x})\|^2 . \tag{4.19}$$

The substitution of $\|\nabla\varphi(\widehat{x})\|^2$ by $\langle \mathscr{B}(\varphi), \varphi \rangle_{\partial B_R}$ in the energy functional has been introduced in [89, 90] for topological derivative evaluation purposes in the context of the domain decomposition method.

Since function w is harmonic in the ball $B_R \subset \mathbb{R}^2$ of radius $R > 0$ and center at $\widehat{x} \in \omega$, then its gradient evaluated at $\widehat{x}$ admits the following representation:

$$\nabla w(\widehat{x}) = \frac{1}{\pi R^3} \int_{\partial B_R} (x - \widehat{x})w(x) . \tag{4.20}$$

Thus, for any $R > \varepsilon$ small enough, the expansion of the energy defined in the ring $C_R(\varepsilon)$ can be written as follows:

$$\int_{C_R(\varepsilon)} \|\nabla w_\varepsilon\|^2 = \int_{B_R} \|\nabla w\|^2$$
$$- \frac{2\varepsilon^2}{\pi R^6} \left[\left(\int_{\partial B_R} w\, x_1 \right)^2 + \left(\int_{\partial B_R} w\, x_2 \right)^2 \right] + o(\varepsilon^2) , \tag{4.21}$$

where $x - \widehat{x} = (x_1, x_2)$. As described in [89, 90], it is interesting to note that the above expansion can be rewritten as

$$\int_{C_R(\varepsilon)} \|\nabla w_\varepsilon\|^2 = \int_{B_R} \|\nabla w\|^2 - \pi\varepsilon^2 \langle \mathscr{B}(w), w \rangle_{\partial B_R} + o(\varepsilon^2) , \tag{4.22}$$

with the nonlocal, positive- and self-adjoint operator $\mathscr{B}$ uniquely determined by the following bilinear form defined on the boundary ∂B_R:

$$\langle \mathscr{B}(w), w \rangle_{\partial B_R} = \frac{2}{\pi^2 R^6} \left[\left(\int_{\partial B_R} w\, x_1 \right)^2 + \left(\int_{\partial B_R} w\, x_2 \right)^2 \right] . \tag{4.23}$$

From the above representation and after taking into account that the line integrals on ∂B_R are well defined by functions in $L^1(\partial B_R)$, the operator $\mathscr{B}$ can be extended to $L^2(\partial B_R)$ as follows:

$$\mathscr{B} \in \mathscr{L}(L^2(\partial B_R); L^2(\partial B_R)) , \tag{4.24}$$

with the same symmetric bilinear form, namely

$$\langle \mathscr{B}(\varphi), \phi \rangle_{\partial B_R} = \frac{2}{\pi^2 R^6} \left[\int_{\partial B_R} \varphi \, x_1 \int_{\partial B_R} \phi \, x_1 + \int_{\partial B_R} \varphi \, x_2 \int_{\partial B_R} \phi \, x_2 \right], \quad (4.25)$$

which is continuous for all $\varphi, \phi \in L^2(\partial B_R)$. It is important to note that the bilinear form

$$L^2(\partial B_R) \times L^2(\partial B_R) \ni (\varphi, \phi) \mapsto \langle \mathscr{B}(\varphi), \phi \rangle_{\partial B_R} \in \mathbb{R} \quad (4.26)$$

is continuous with respect to the weak convergence since it has a simple structure, namely

$$\langle \mathscr{B}(\varphi), \phi \rangle_{\partial B_R} = L_1(\varphi) L_1(\phi) + L_2(\varphi) L_2(\phi) \quad \varphi, \phi \in L^1(\partial B_R), \quad (4.27)$$

with two linear forms $\varphi \mapsto L_1(\varphi)$ and $\phi \mapsto L_2(\phi)$ given by line integrals on ∂B_R. Finally, since the asymptotic expansion of the Steklov–Poincaré operator coincides with the asymptotic expansion of the energy in the ring $C_R(\varepsilon)$, we have

$$\langle \mathscr{A}_\varepsilon(\varphi), \varphi \rangle_{\partial B_R} = \langle \mathscr{A}(\varphi), \varphi \rangle_{\partial B_R} - f(\varepsilon) \langle \mathscr{B}(\varphi), \varphi \rangle_{\partial B_R} + o(f(\varepsilon)), \quad (4.28)$$

with $f(\varepsilon) = \pi \varepsilon^2$. Therefore, the asymptotic expansion of the energy shape functional in $C_R(\varepsilon)$ for $\varepsilon \to 0$ is given by the regular expansion of the *Steklov–Poincaré operator* (see Sect. 4.2.1, Lemma 4.1):

$$\mathscr{A}_\varepsilon = \mathscr{A} - f(\varepsilon) \mathscr{B} + \mathscr{R}_\varepsilon, \quad (4.29)$$

where the remainder term denoted as $\mathscr{R}_\varepsilon$ in the above expansion is of order $o(f(\varepsilon))$ in the operator norm $\mathscr{L}(H^{1/2}(\partial B_R); H^{-1/2}(\partial B_R))$. From the symmetry of the operators, the expansion of the energy shape functional can also be written as

$$\langle \mathscr{A}_\varepsilon(\varphi), \phi \rangle_{\partial B_R} = \langle \mathscr{A}(\varphi), \phi \rangle_{\partial B_R} - f(\varepsilon) \langle \mathscr{B}(\varphi), \phi \rangle_{\partial B_R} + \langle \mathscr{R}_\varepsilon(\varphi), \phi \rangle_{\partial B_R}, \quad (4.30)$$

where $\langle \mathscr{R}_\varepsilon(\varphi), \phi \rangle_{\partial B_R} = o(f(\varepsilon))$. In view of the asymptotic expansion of the energy shape functional, the representation (4.25) holds true or alternatively

$$\langle \mathscr{B}(\varphi), \phi \rangle_{\partial B_R} = 2 \nabla \varphi(\widehat{x}) \cdot \nabla \phi(\widehat{x}) \quad \forall \widehat{x} \in \omega, \quad (4.31)$$

provided that ϕ and φ are analytic. Note that, for the sake of notation, the functions ϕ and φ are not distinguished from their respective traces evaluated on the fictitious boundary ∂B_R.

4.2.1 Compactness of the Asymptotic Expansion

Thanks to the domain decomposition technique, the compactness of the obtained asymptotic expansion is ensured in this section by using elementary arguments from the *Fourier analysis*. Let φ be a function with trace (still denoted as φ) on ∂B_R belonging to $H^{1/2}(\partial B_R)$, then

$$\|\varphi\|_{H^{1/2}(\partial B_R)} \leq C_R . \tag{4.32}$$

By taking into account that the radius R is fixed, for the sake of simplicity the subindex R will be omitted in what follows, so that C_R is replaced by C, with C used to denote a generic constant independent of the small parameter ε. Finally, let us introduce a polar coordinate system (r, θ) centered at $\widehat{x}$. Since $\varphi \in H^{1/2}(\partial B_R)$, it follows that there exists an expansion in *Fourier series* of φ in terms of θ of the form:

$$\varphi(\theta) = \frac{1}{2}a_0 + \sum_{k=1}^{\infty}(a_k \sin k\theta + b_k \cos k\theta) , \tag{4.33}$$

with the associated coefficients satisfying

$$\sum_{k=1}^{\infty} \sqrt{1 + k^2}(a_k^2 + b_k^2) \leq C . \tag{4.34}$$

From these elements, two important properties are derived, namely

$$\sum_{k=1}^{\infty}(a_k^2 + b_k^2) \leq C \quad \text{and} \quad \sum_{k=1}^{\infty} k(a_k^2 + b_k^2) \leq C . \tag{4.35}$$

Let us consider the solution w of the Laplace equation defined in the ball B_R endowed with Dirichlet boundary condition on ∂B_R given by φ. Let us also consider the solution w_ε of the same problem but defined in the ring $C_R(\varepsilon) = B_R \setminus \overline{B_\varepsilon}$ endowed with homogeneous Neumann boundary condition on ∂B_ε. Then, the associated energies depending on φ through the boundary conditions on ∂B_R are given respectively by

$$\mathscr{E}(\varphi) = \int_{\partial B_R} \mathscr{A}(\varphi)\varphi \equiv \int_{B_R} \|\nabla w\|^2 , \tag{4.36}$$

$$\mathscr{E}_\varepsilon(\varphi) = \int_{\partial B_R} \mathscr{A}_\varepsilon(\varphi)\varphi \equiv \int_{C_R(\varepsilon)} \|\nabla w_\varepsilon\|^2 . \tag{4.37}$$

We need to show that $\mathscr{E}_\varepsilon(\varphi)$ has an asymptotic expansion with respect to ε whose remainder term is uniformly bounded. More precisely, the following result has to be justified:

Lemma 4.1 *The energy $\mathscr{E}_\varepsilon(\varphi)$ admits and asymptotic expansion for $\varepsilon > 0$, ε small enough, of the form*

$$\mathscr{E}_\varepsilon(\varphi) = \mathscr{E}(\varphi) - \pi\varepsilon^2 \langle \mathscr{B}(\varphi), \varphi \rangle_{\partial B_R} + \langle \mathscr{R}_\varepsilon(\varphi), \varphi \rangle_{\partial B_R} , \tag{4.38}$$

with

$$|\langle \mathscr{R}_\varepsilon(\varphi), \varphi \rangle_{\partial B_R}| \leq C\varepsilon^4 \tag{4.39}$$

uniformly for any fixed compact set in $H^1(\mathscr{D} \setminus \overline{B_R})$, i.e., C depends on this set only.

Proof By taking into account that all compact set can be covered by a finite number of balls, it is sufficient to show the result for one single fixed ball B_R. Therefore, we can assume that (4.35) holds true. The proof consists in deriving the *explicit formulas* for w and w_ε written in terms of Fourier series, similar to [88]. Thus, the associated energies can be evaluated explicitly and the properties of the remainder term $\mathscr{R}_\varepsilon(\varphi)$ immediately deduced. By constructing w from an expansion in *Fourier series* of the boundary condition on ∂B_R, we have

$$w(r, \theta) = \frac{1}{2}a_0 + \sum_{k=1}^{\infty} \left(\frac{r}{R}\right)^k (a_k \sin k\theta + b_k \cos k\theta) . \tag{4.40}$$

Similarly for w_ε in $C_R(\varepsilon)$, the following expansion in *Fourier series* is valid:

$$w_\varepsilon(r, \theta) = \frac{1}{2}a_0 + \sum_{k=1}^{\infty} \psi_k(r)(a_k \sin k\theta + b_k \cos k\theta) , \tag{4.41}$$

where

$$\psi_k(r) = A_k r^k + B_k r^{-k} , \tag{4.42}$$

with A_k and B_k determined by the boundary conditions on ∂B_R and ∂B_ε, that is

$$A_k R^k + B_k \frac{1}{R^k} = 1 \quad \text{and} \quad A_k \varepsilon^{k-1} - B_k \frac{1}{\varepsilon^{k+1}} = 0 . \tag{4.43}$$

Therefore

$$A_k = \frac{R^k}{R^{2k} + \varepsilon^{2k}} \quad \text{and} \quad B_k = A_k \varepsilon^{2k} , \tag{4.44}$$

and finally

$$\psi_k(r) = \frac{r^k}{R^k} + \frac{\varepsilon^{2k}}{R^{2k} + \varepsilon^{2k}} \left(\frac{R^k}{r^k} - \frac{r^k}{R^k} \right) . \tag{4.45}$$

After replacing this last result in the expansion for w_ε, we get

$$w_\varepsilon(r, \theta) = w(r, \theta) + \widetilde{w}_\varepsilon(r, \theta) , \tag{4.46}$$

with

$$\widetilde{w}_\varepsilon(r, \theta) = \sum_{k=1}^{\infty} \frac{\varepsilon^{2k}}{R^{2k} + \varepsilon^{2k}} \left(\frac{R^k}{r^k} - \frac{r^k}{R^k} \right) (a_k \sin k\theta + b_k \cos k\theta) . \tag{4.47}$$

Thus,

$$\begin{aligned} \mathscr{E}_\varepsilon(\varphi) &= \int_{C_R(\varepsilon)} \|\nabla w + \nabla \widetilde{w}_\varepsilon\|^2 \\ &= \int_{C_R(\varepsilon)} \|\nabla w\|^2 + 2 \int_{C_R(\varepsilon)} \nabla w \cdot \nabla \widetilde{w}_\varepsilon + \int_{C_R(\varepsilon)} \|\nabla \widetilde{w}_\varepsilon\|^2 \pm \int_{B_\varepsilon} \|\nabla w\|^2 \\ &= \mathscr{E}(\varphi) + \mathscr{I}_1 + \mathscr{I}_2 + \mathscr{I}_3 , \end{aligned} \tag{4.48}$$

where the integrals $\mathscr{I}_1$, $\mathscr{I}_2$ and $\mathscr{I}_3$ are defined as

$$\mathscr{I}_1 := \int_{C_R(\varepsilon)} \|\nabla \widetilde{w}_\varepsilon\|^2, \quad \mathscr{I}_2 := 2 \int_{C_R(\varepsilon)} \nabla w \cdot \nabla \widetilde{w}_\varepsilon \text{ and } \mathscr{I}_3 := - \int_{B_\varepsilon} \|\nabla w\|^2. \tag{4.49}$$

Now we have

$$\partial_r \widetilde{w}_\varepsilon(r, \theta) = - \sum_{k=1}^{\infty} \frac{\varepsilon^{2k}}{R^{2k} + \varepsilon^{2k}} k \frac{1}{r} \left(\frac{R^k}{r^k} + \frac{r^k}{R^k} \right) (a_k \sin k\theta + b_k \cos k\theta) , \tag{4.50}$$

$$\frac{1}{r} \partial_\theta \widetilde{w}_\varepsilon(r, \theta) = \sum_{k=1}^{\infty} \frac{\varepsilon^{2k}}{R^{2k} + \varepsilon^{2k}} k \frac{1}{r} \left(\frac{R^k}{r^k} - \frac{r^k}{R^k} \right) (a_k \cos k\theta - b_k \sin k\theta) . \tag{4.51}$$

After integrating with respect to θ and in view of the orthogonality of the trigonometric functions between 0 and 2π, we have

$$\mathscr{I}_1 = \pi \sum_{k=1}^{\infty} \left(\frac{\varepsilon^{2k}}{R^{2k} + \varepsilon^{2k}} \right)^2 k^2 (a_k^2 + b_k^2) \mathscr{I}_k(\varepsilon) , \tag{4.52}$$

where the integral $\mathscr{I}_k(\varepsilon)$ is defined as

$$\mathscr{I}_k(\varepsilon) = \int_\varepsilon^R \left[\left(\frac{R^k}{r^{k+1}} + \frac{r^{k-1}}{R^k} \right)^2 + \left(\frac{R^k}{r^{k+1}} - \frac{r^{k-1}}{R^k} \right)^2 \right]$$

$$r\, dr = \frac{1}{k} \left(\frac{R^{2k}}{\varepsilon^{2k}} - \frac{\varepsilon^{2k}}{R^{2k}} \right), \tag{4.53}$$

which leads to

$$\mathscr{I}_1 = \pi \sum_{k=1}^\infty \left(\frac{\varepsilon^{2k}}{R^{2k} + \varepsilon^{2k}} \right)^2 k(a_k^2 + b_k^2) \left(\frac{R^{2k}}{\varepsilon^{2k}} - \frac{\varepsilon^{2k}}{R^{2k}} \right)$$

$$= \pi \varepsilon^2 \frac{a_1^2 + b_1^2}{R^2} + O(\varepsilon^4) . \tag{4.54}$$

Before evaluating the second integral $\mathscr{I}_2$, we observe that

$$\partial_r w(r, \theta) = \sum_{k=1}^\infty k \frac{r^{k-1}}{R^k} (a_k \sin k\theta + b_k \cos k\theta) , \tag{4.55}$$

$$\frac{1}{r} \partial_\theta w(r, \theta) = \sum_{k=1}^\infty k \frac{r^{k-1}}{R^k} (a_k \cos k\theta - b_k \sin k\theta) . \tag{4.56}$$

A simple manipulation yields

$$\mathscr{I}_2 = 2\pi \sum_{k=1}^\infty \left(\frac{\varepsilon}{R} \right)^{2k} k(a_k^2 + b_k^2) \frac{R^{2k} - \varepsilon^{2k}}{R^{2k} + \varepsilon^{2k}} = -2\pi \varepsilon^2 \frac{a_1^2 + b_1^2}{R^2} + O(\varepsilon^4) .$$
$$\tag{4.57}$$

After evaluating the last integral $\mathscr{I}_3$, we obtain

$$\mathscr{I}_3 = -\pi \sum_{k=1}^\infty \left(\frac{\varepsilon}{R} \right)^{2k} k(a_k^2 + b_k^2) = -\pi \varepsilon^2 \frac{a_1^2 + b_1^2}{R^2} + O(\varepsilon^4) . \tag{4.58}$$

Finally, from the obtained formulas (4.54), (4.57), and (4.58), it is possible to identify the term independent of ε as well as the term of order ε^2, that is

$$\mathscr{E}_\varepsilon(\varphi) = \mathscr{E}(\varphi) - 2\pi \varepsilon^2 \frac{a_1^2 + b_1^2}{R^2} + O(\varepsilon^4) , \tag{4.59}$$

where, in view of the inequalities (4.35), the remainder of order $O(\varepsilon^4)$ is uniformly bounded by $C\varepsilon^4$. □

4.2.2 Asymptotic Expansion of the Solution

Let us consider an *ansatz* for the solution u_ε to the topologically perturbed coupled problem (4.7) of the form

$$u_\varepsilon(x) = u(x) + f(\varepsilon)g(x) + \tilde{u}_\varepsilon(x) , \tag{4.60}$$

where u is the solution to the original (unperturbed) coupled problem (4.2), g is the first order asymptotic corrector, and $\tilde{u}_\varepsilon$ is the *remainder*. We need to expand the application $\mathscr{A}_\varepsilon(u_\varepsilon)$ in power of ε. For that, we just introduce the ansatz (4.60) into the expansion of the Steklov–Poincaré operator (4.29) to obtain

$$\mathscr{A}_\varepsilon(u_\varepsilon) = \mathscr{A}(u) + f(\varepsilon)(\mathscr{A}(g) + \mathscr{B}(u)) + \mathscr{A}_\varepsilon(\tilde{u}_\varepsilon)$$
$$+ \mathscr{R}_\varepsilon(u) + f(\varepsilon)\mathscr{R}_\varepsilon(g) + f(\varepsilon)^2 \mathscr{B}(g) \quad \text{on} \quad \partial B_R . \tag{4.61}$$

By considering these last two results in (4.12) and after collecting the terms in power of ε, we can define one variational problem for each term in the ansatz (4.60). The first problem for u is given by: Find $u \in \mathscr{V}_R$, such that

$$\int_{\omega \backslash \overline{B_R}} \nabla u \cdot \nabla \eta + \int_{\partial B_R} \mathscr{A}(u)\eta + \int_\Omega \nabla u \cdot \nabla \eta + \int_\Omega u\eta = \int_\Omega b\eta \quad \forall \eta \in \mathscr{V}_R . \tag{4.62}$$

The second problem for g is stated as follows: Find $g \in \mathscr{V}_R$, such that

$$\int_{\omega \backslash \overline{B_R}} \nabla g \cdot \nabla \eta + \int_{\partial B_R} \mathscr{A}(g)\eta + \int_\Omega \nabla g \cdot \nabla \eta + \int_\Omega g\eta = - \int_{\partial B_R} \mathscr{B}(u)\eta \quad \forall \eta \in \mathscr{V}_R . \tag{4.63}$$

Finally, the third problem for the reminder $\tilde{u}_\varepsilon$ is written as: Find $\tilde{u}_\varepsilon \in \mathscr{V}_R$, such that

$$\int_{\omega \backslash \overline{B_R}} \nabla \tilde{u}_\varepsilon \cdot \nabla \eta + \int_{\partial B_R} \mathscr{A}_\varepsilon(\tilde{u}_\varepsilon)\eta + \int_\Omega \nabla \tilde{u}_\varepsilon \cdot \nabla \eta + \int_\Omega \tilde{u}_\varepsilon \eta = \int_{\partial B_R} \mathfrak{F}_\varepsilon \eta \quad \forall \eta \in \mathscr{V}_R , \tag{4.64}$$

where the source term $\mathfrak{F}_\varepsilon$ is given by

$$\mathfrak{F}_\varepsilon = -(\mathscr{R}_\varepsilon(u) + f(\varepsilon)\mathscr{R}_\varepsilon(g) + f(\varepsilon)^2 \mathscr{B}(g)) . \tag{4.65}$$

The estimate $\|\tilde{u}_\varepsilon\|_{H^1(\mathscr{D} \backslash \overline{B_R})} = o(f(\varepsilon))$ for the remainder holds true, whose proof is left as an exercise at the end of this chapter.

4.2.3 Asymptotic Expansion of the Shape Functional

Now, we have to derive the asymptotic expansion of the shape functional and find the corresponding topological derivative. After introducing the ansatz (4.60) in the shape functional regarding the topologically perturbed problem (4.6), we have

$$\mathscr{J}_\Omega(u_\varepsilon) = \frac{1}{2} \int_\Omega |u + f(\varepsilon)g + \tilde{u}_\varepsilon - z_d|^2$$

$$= \frac{1}{2} \int_\Omega |u - z_d|^2 + f(\varepsilon) \int_\Omega (u - z_d)g + o(f(\varepsilon)) . \qquad (4.66)$$

Let us rewrite the adjoint problem (4.5) as follows: Find $v \in \mathscr{V}_R$, such that

$$\int_{\omega \setminus \overline{B_R}} \nabla v \cdot \nabla \eta + \int_{\partial B_R} \mathscr{A}(v)\eta + \int_\Omega \nabla v \cdot \nabla \eta + \int_\Omega v\eta = -\int_\Omega (u - z_d)\eta \quad \forall \eta \in \mathscr{V}_R .$$

$$(4.67)$$

By taking g as test function in the above variational problem we have

$$\int_{\omega \setminus \overline{B_R}} \nabla v \cdot \nabla g + \int_{\partial B_R} \mathscr{A}(v)g + \int_\Omega \nabla v \cdot \nabla g + \int_\Omega vg = -\int_\Omega (u - z_d)g . \qquad (4.68)$$

On the other hand, by setting v as test function in the variational problem given by (4.63), there is

$$\int_{\omega \setminus \overline{B_R}} \nabla g \cdot \nabla v + \int_{\partial B_R} \mathscr{A}(g)v + \int_\Omega \nabla g \cdot \nabla v + \int_\Omega gv = -\int_{\partial B_R} \mathscr{B}(u)v . \qquad (4.69)$$

After combining both equalities, the following important result yields:

$$\int_\Omega (u - z_d)g = \int_{\partial B_R} \mathscr{B}(u)v = \langle \mathscr{B}(u), v \rangle_{\partial B_R} = 2\nabla u(\hat{x}) \cdot \nabla v(\hat{x}) , \qquad (4.70)$$

where we have considered the symmetry of the bilinear forms together with the representation (4.31), which is well defined, thanks to the interior elliptic regularity of u and v. Finally, after introducing this last result in (4.66), the *topological asymptotic expansion* of the shape functional leads to

$$\psi(\chi_\varepsilon(\hat{x})) = \psi(\chi) + 2f(\varepsilon)\nabla u(\hat{x}) \cdot \nabla v(\hat{x}) + o(f(\varepsilon)) . \qquad (4.71)$$

Therefore, the associated *topological derivative* can be promptly identified, namely

$$\mathscr{T}(\hat{x}) = 2\nabla u(\hat{x}) \cdot \nabla v(\hat{x}) \quad \forall \hat{x} \in \omega , \qquad (4.72)$$

with function $f(\varepsilon) = \pi\varepsilon^2$. Let us stress that u and v are respectively solutions to the direct (4.2) and adjoint (4.5) problems, both defined in the original (unperturbed) domain $\mathscr{D}$.

4.3 Exercises

1. Show that the remainder $\widetilde{u}_\varepsilon$, satisfying (4.64), enjoys the property

$$\|\widetilde{u}_\varepsilon\|_{H^1(\mathscr{D}\setminus\overline{B_R})} = o(f(\varepsilon)) \ .$$

   Hint: Study the proof of Lemma 3.2, Chap. 3, and adapt it for the case analyzed here.
2. Repeat the analysis presented in this chapter by considering the following elliptic nonlinear coupled problem: Find $u \in H_0^1(\mathscr{D})$, such that

$$\int_\omega \nabla u \cdot \nabla\eta + \int_\Omega \nabla u \cdot \nabla\eta + \int_\Omega u^3\eta = \int_\Omega b\eta \quad \forall\eta \in H_0^1(\mathscr{D}) \ .$$

   Hint: Take a look on the book by Novotny and Sokołowski [75, Chapters 10 and 11].

Chapter 5
Topology Design Optimization

In this chapter a topology optimization algorithm based on the topological derivative concept combined with a level-set domain representation method is presented [11], together with its applications in the context of compliance structural topology optimization and topology design of compliant mechanisms. It is worth mentioning that the topological derivative is defined through a limit passage when the small parameter governing the size of the topological perturbation goes to zero. Therefore, it can be used as a steepest-descent direction in an optimization process, according to any method based on the gradient of the cost functional. We restrict ourselves to the case in which the domain is topologically perturbed by the nucleation of a small inclusion where a weak material phase is used to mimic voids, allowing to work in a fixed computational domain. This simple strategy bypasses the use of a complicated algorithm specifically designed to deal with nucleation of holes in a computational domain.

Let us introduce a hold-all domain $\mathscr{D} \subset \mathbb{R}^2$, which is split into two subdomains, $\Omega \subset \mathscr{D}$ and its complement $\mathscr{D} \setminus \Omega$. We assume that there is a distributed parameter $\rho : \mathscr{D} \mapsto \{1, \rho_0\}$ defined as

$$\rho(x) := \begin{cases} 1 & \text{if } x \in \Omega, \\ \rho_0 & \text{if } x \in \mathscr{D} \setminus \Omega, \end{cases} \tag{5.1}$$

with $0 < \rho_0 \ll 1$. The topology optimization problem we are dealing with consists in minimizing a shape functional $\Omega \mapsto J(\Omega)$ with respect to $\Omega \subset \mathscr{D}$, that is:

$$\underset{\Omega \subset \mathscr{D}}{\text{Minimize }} J(\Omega) , \tag{5.2}$$

which can be solved by using the topological derivative concept. Actually, a circular hole $B_\varepsilon(\widehat{x})$ is introduced inside $\mathscr{D}$. Then, the region occupied by $B_\varepsilon(\widehat{x})$ is filled by

A. A. Novotny, J. Sokołowski, *An Introduction to the Topological Derivative Method*, SpringerBriefs in Mathematics, https://doi.org/10.1007/978-3-030-36915-6_5

an inclusion with material properties different from the background. The material properties are characterized by a piecewise constant function γ_ε of the form

$$\gamma_\varepsilon(x) := \begin{cases} 1 & \text{if } x \in \mathcal{D} \setminus \overline{B_\varepsilon}, \\ \gamma(x) & \text{if } x \in B_\varepsilon, \end{cases} \tag{5.3}$$

where the *contrast* γ is defined as

$$\gamma(x) = \begin{cases} \rho_0 & \text{if } x \in \Omega, \\ \rho_0^{-1} & \text{if } x \in \mathcal{D} \setminus \Omega, \end{cases} \tag{5.4}$$

which induces a level-set domain representation method.

In order to fix these ideas, a model problem in elasticity is considered in Sect. 5.1. The resulting topology design algorithm based on the topological derivative concept combined with a level-set domain representation method is presented in Sect. 5.2. Some numerical results in the context of compliance structural topology optimization and topology design of compliant mechanisms are presented in Sect. 5.3. Finally, the chapter ends in Sect. 5.4 with a discussion concerning perspectives of future developments, together with a list of open problems.

5.1 Model Problem in Elasticity

In this section, the topological derivative of a tracking-type shape functional associated with the linear elasticity problem into two spatial dimensions, in the presence of an small circular inclusion, is derived.

The tracking-type shape functional associated with the unperturbed domain is defined as

$$\psi(\chi) := \mathscr{J}(u) = \int_{\Gamma_N} g \cdot u, \tag{5.5}$$

where g is a given vector function in $H^{-1/2}(\Gamma_N)$ and the displacement vector field $u : \mathcal{D} \mapsto \mathbb{R}^2$ is the solution of the following variational problem:

$$u \in \mathscr{U} : \int_{\mathcal{D}} \sigma(u) \cdot (\nabla \eta)^s = \int_{\Gamma_N} q \cdot \eta \quad \forall \eta \in \mathscr{V}, \tag{5.6}$$

with $\sigma(u) = \rho\mathbb{C}(\nabla u)^s$. In the above equation, ρ is given by (5.1), $q \in H^{-1/2}(\Gamma_N)$ is a given boundary traction, and $(\nabla\varphi)^s$ is the symmetric part of the gradient of a vector field φ, namely

$$(\nabla\varphi)^s := \frac{1}{2}\left(\nabla\varphi + (\nabla\varphi)^\top\right). \tag{5.7}$$

By considering isotropic medium, the constitutive tensor $\mathbb{C}$ can be represented as

$$\mathbb{C} = 2\mu\,\mathbb{I} + \lambda\,\mathrm{I} \otimes \mathrm{I}\,, \tag{5.8}$$

where I and $\mathbb{I}$ are identity tensors of second and fourth orders, respectively, and μ and λ are the Lamé coefficients, both considered constants everywhere. In particular, in the case of plane stress assumptions, we have

$$\mu = \frac{E}{2(1+\nu)} \quad \text{and} \quad \lambda = \frac{\nu E}{1-\nu^2}\,, \tag{5.9}$$

whereas in the case of plane strain state, there are

$$\mu = \frac{E}{2(1+\nu)} \quad \text{and} \quad \lambda = \frac{\nu E}{(1+\nu)(1-2\nu)}\,, \tag{5.10}$$

where E is the Young modulus and ν the Poisson ratio. The spaces $\mathscr{U}$ and $\mathscr{V}$ are defined as

$$\mathscr{U} = \mathscr{V} = \left\{ \varphi \in H^1(\mathscr{D}) : \varphi|_{\Gamma_D} = 0 \right\}. \tag{5.11}$$

In addition, $\partial\mathscr{D} = \Gamma_D \cup \Gamma_N$ with $\Gamma_D \cap \Gamma_N = \varnothing$, where Γ_D and Γ_N are Dirichlet and Neumann boundaries, respectively. See sketch in Fig. 5.1.

The strong system associated with the variational problem (5.6) can be stated as: Find u, such that

$$\begin{cases} \operatorname{div}\sigma(u) = 0 & \text{in } \mathscr{D}\,, \\ \sigma(u) = \rho\mathbb{C}(\nabla u)^s & \\ u = 0 & \text{on } \Gamma_D\,, \\ \sigma(u)n = q & \text{on } \Gamma_N\,. \end{cases} \tag{5.12}$$

Remark 5.1 By setting $\rho(x) = 1\ \forall x \in \mathscr{D}$ in (5.1), the boundary value problem (5.12) degenerates itself to the well-known Navier system, namely

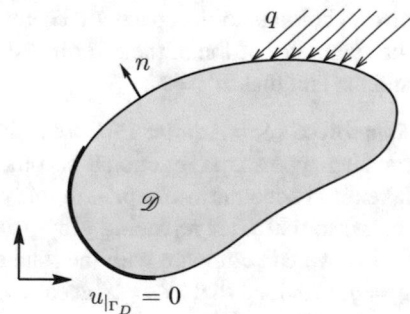

Fig. 5.1 The elasticity problem defined in the unperturbed domain

$$\mu \Delta u + (\lambda + \mu) \nabla (\operatorname{div} u) = 0 \quad \text{in} \quad \mathscr{D} , \tag{5.13}$$

where μ and λ are the Lamé coefficients given by (5.9) for the plane stress case and by (5.10) for the plane strain assumption.

In order to simplify further analysis, an auxiliary vector function $v : \mathscr{D} \mapsto \mathbb{R}^2$ is introduced, which is the solution of the following *adjoint variation problem* (see Sect. 1.2.1 for details)

$$v \in \mathscr{V} : \int_{\mathscr{D}} \sigma(v) \cdot (\nabla \eta)^s = - \int_{\Gamma_N} g \cdot \eta \quad \forall \eta \in \mathscr{V} , \tag{5.14}$$

with $\sigma(v) = \rho \mathbb{C} (\nabla v)^s$.

Remark 5.2 (Lagrangian Formalism) As discussed in Sect. 1.2.1, the adjoint state v solution of (5.14) comes out from the Lagrangian formalism. In particular, the basic idea consists in defining a Lagrangian functional given by the sum of the shape functional (5.5) and the state equation in its weak form (5.6), namely

$$\mathscr{L}(u, v) := \int_{\Gamma_N} g \cdot u + \int_{\mathscr{D}} \sigma(u) \cdot (\nabla v)^s - \int_{\Gamma_N} q \cdot v . \tag{5.15}$$

By applying the first order optimality condition in (5.15) with respect to $v \in \mathscr{V}$, we recover the state equation (5.6), that is

$$u \in \mathscr{U} : \int_{\mathscr{D}} \sigma(u) \cdot (\nabla \eta)^s - \int_{\Gamma_N} q \cdot \eta = 0 \quad \forall \eta \in \mathscr{V} . \tag{5.16}$$

On the other hand, after applying the first order optimality condition in (5.15) with respect to $u \in \mathscr{U}$, we obtain

$$v \in \mathscr{V} : \int_{\mathscr{D}} \sigma(\eta) \cdot (\nabla v)^s + \int_{\Gamma_N} g \cdot \eta = 0 \quad \forall \eta \in \mathscr{V} , \tag{5.17}$$

which is actually the adjoint equation (5.14), since the bilinear form on the left-hand side of (5.17) is symmetric. It is also important to note that the adjoint state v always belongs to the space $\mathscr{V}$. Therefore, in our particular case, we have just used the symmetry of the bilinear form to define the adjoint problem according to (5.14) and the fact that $\mathscr{U} = \mathscr{V}$.

Remark 5.3 (Self Adjoint Problem) Note also that by setting $g = q$ in (5.5), the tracking-type shape functional becomes the structural compliance, which has been taken into account in Chap. 3. In this particular case, the problem is self-adjoint in the sense that after replacing g by q in the right-hand side of the adjoint equation (5.14), we can compare with the state equation (5.6) and conclude that $v = -u$ for $g = q$, provided that $\mathscr{U} = \mathscr{V}$ according to (5.11).

Now, let us state the perturbed counterpart of the problem. In particular, the tracking-type shape functional associated with the topologically perturbed domain can be written as

$$\psi(\chi_\varepsilon) := \mathcal{J}_\varepsilon(u_\varepsilon) = \int_{\Gamma_N} g \cdot u_\varepsilon \,. \tag{5.18}$$

The displacement vector field $u_\varepsilon : \mathcal{D} \mapsto \mathbb{R}^2$ solves the following variational problem:

$$u_\varepsilon \in \mathcal{U} : \int_{\mathcal{D}} \sigma_\varepsilon(u_\varepsilon) \cdot (\nabla\eta)^s = \int_{\Gamma_N} q \cdot \eta \quad \forall \eta \in \mathcal{V} \,, \tag{5.19}$$

with $\sigma_\varepsilon(u_\varepsilon) = \gamma_\varepsilon \rho \mathbb{C}(\nabla u_\varepsilon)^s$, where the contrast γ_ε is given by (5.3). The *strong system* associated with the variational problem (5.19) can be written as: Find u_ε, such that

$$\begin{cases} \operatorname{div} \sigma_\varepsilon(u_\varepsilon) = 0 & \text{in } \mathcal{D} \,, \\ \sigma_\varepsilon(u_\varepsilon) = \gamma_\varepsilon \rho \mathbb{C}(\nabla u_\varepsilon)^s & \\ u_\varepsilon = 0 & \text{on } \Gamma_D \,, \\ \sigma(u_\varepsilon)n = q & \text{on } \Gamma_N \,, \\ \left.\begin{array}{l} [\![u_\varepsilon]\!] = 0 \\ [\![\sigma_\varepsilon(u_\varepsilon)]\!]n = 0 \end{array}\right\} & \text{on } \partial B_\varepsilon \,, \end{cases} \tag{5.20}$$

where the operator $[\![\varphi]\!]$ is used to denote the jump of function φ on the boundary of the inclusion ∂B_ε, namely $[\![\varphi]\!] := \varphi|_{\mathcal{D}\setminus\overline{B_\varepsilon}} - \varphi|_{B_\varepsilon}$ on ∂B_ε. See details in Fig. 5.2. Note that the *transmission condition* on the interface ∂B_ε comes out from the variational formulation (5.19).

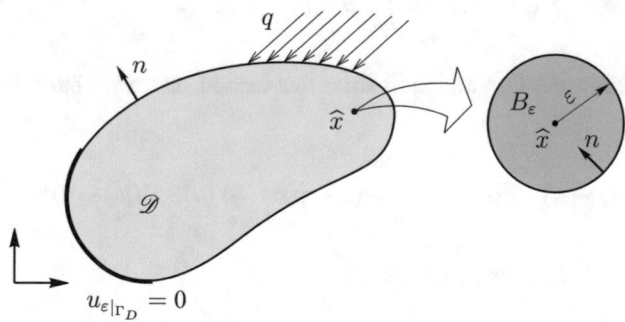

$$u_\varepsilon|_{\Gamma_D} = 0$$

Fig. 5.2 The elasticity problem defined in the perturbed domain

5.1.1 Existence of the Topological Derivative

The following lemma ensures the existence of the associated topological derivative:

Lemma 5.1 *Let u and u_ε be the solutions of the original* (5.6) *and perturbed* (5.19) *problems, respectively. Then, the following estimate holds true:*

$$\|u_\varepsilon - u\|_{H^1(\mathscr{D})} \leq C\varepsilon ,\tag{5.21}$$

where C is a constant independent of the control parameter ε.

Proof From the definition of the contrast γ_ε given by (5.3), we have that Eq. (5.6) can be rewritten as

$$\int_{\mathscr{D}\backslash \overline{B_\varepsilon}} \sigma(u)\cdot(\nabla\eta)^s + \int_{B_\varepsilon} \sigma(u)\cdot(\nabla\eta)^s \pm \int_{B_\varepsilon} \gamma\sigma(u)\cdot(\nabla\eta)^s = \int_{\Gamma_N} q\cdot\eta ,\tag{5.22}$$

or even as

$$u \in \mathscr{U} \; : \; \int_{\mathscr{D}} \sigma_\varepsilon(u)\cdot(\nabla\eta)^s + (1-\gamma)\int_{B_\varepsilon} \sigma(u)\cdot(\nabla\eta)^s = \int_{\Gamma_N} \overline{q}\cdot\eta \quad \forall\eta\in\mathscr{V} .\tag{5.23}$$

By taking $\eta = u_\varepsilon - u$ as test function in the above equation and also in (5.19), there are

$$\int_{\mathscr{D}} \sigma_\varepsilon(u)\cdot\nabla(u_\varepsilon-u)^s = \int_{\Gamma_N} q\cdot(u_\varepsilon-u)$$

$$-(1-\gamma)\int_{B_\varepsilon} \sigma(u)\cdot\nabla(u_\varepsilon-u)^s ,\tag{5.24}$$

$$\int_{\mathscr{D}} \sigma_\varepsilon(u_\varepsilon)\cdot\nabla(u_\varepsilon-u)^s = \int_{\Gamma_N} q\cdot(u_\varepsilon-u) .\tag{5.25}$$

After subtracting the first equation from the second one, we obtain the following equality:

$$\int_{\mathscr{D}} \sigma_\varepsilon(u_\varepsilon-u)\cdot\nabla(u_\varepsilon-u)^s = (1-\gamma)\int_{B_\varepsilon} \sigma(u)\cdot\nabla(u_\varepsilon-u)^s .\tag{5.26}$$

The Cauchy–Schwarz inequality implies

$$\int_{\mathscr{D}} \sigma_\varepsilon(u_\varepsilon-u)\cdot\nabla(u_\varepsilon-u)^s \leq C_1\|\sigma(u)\|_{L^2(B_\varepsilon)}\|\nabla(u_\varepsilon-u)^s\|_{L^2(B_\varepsilon)}$$

$$\leq C_2\varepsilon\|u_\varepsilon-u\|_{H^1(\mathscr{D})} ,\tag{5.27}$$

where we have used the interior elliptic regularity of u. Finally, from the *coercivity* of the bilinear form on the left-hand side of the above inequality, namely

$$c\|u_\varepsilon - u\|^2_{H^1(\mathscr{D})} \leq \int_{\mathscr{D}} \sigma_\varepsilon(u_\varepsilon - u) \cdot \nabla(u_\varepsilon - u)^s , \tag{5.28}$$

we obtain

$$\|u_\varepsilon - u\|^2_{H^1(\mathscr{D})} \leq C\varepsilon\|u_\varepsilon - u\|_{H^1(\mathscr{D})} , \tag{5.29}$$

which leads to the result with $C = C_2/c$.   □

5.1.2 Variation of the Shape Functional

From a simple manipulation and with the help of the adjoint equation (5.14), it is possible to write the variation of the shape functional in terms of an integral concentrated in the ball B_ε. In fact, after subtracting (5.5) from (5.18) we obtain

$$\mathscr{J}_\varepsilon(u_\varepsilon) - \mathscr{J}(u) = \int_{\Gamma_N} g \cdot (u_\varepsilon - u) . \tag{5.30}$$

From the definition for the contrast γ_ε given by (5.3), the state equation associated with the topologically perturbed domain (5.19) can be rewritten as

$$\int_{\mathscr{D}\backslash\overline{B_\varepsilon}} \sigma(u_c) \cdot (\nabla\eta)^s + \int_{B_\varepsilon} \gamma\sigma(u_\varepsilon) \cdot (\nabla\eta)^s \perp \int_{B_\varepsilon} \upsilon(u_\varepsilon) \cdot (\nabla\eta)^s = \int_{\Gamma_N} q \cdot \eta . \tag{5.31}$$

Therefore, it follows that

$$\int_{\mathscr{D}} \sigma(u_\varepsilon) \cdot (\nabla\eta)^s = (1 - \gamma)\int_{B_\varepsilon} \sigma(u_\varepsilon) \cdot (\nabla\eta)^s + \int_{\Gamma_N} q \cdot \eta . \tag{5.32}$$

Now, we can subtract the state equation associated with the unperturbed domain (5.6) from the above result to obtain

$$\int_{\mathscr{D}} \sigma(u_\varepsilon - u) \cdot (\nabla\eta)^s = (1 - \gamma)\int_{B_\varepsilon} \sigma(u_\varepsilon) \cdot (\nabla\eta)^s . \tag{5.33}$$

By choosing $\eta = v$ as test function in the above equation, where v is the adjoint state solution of (5.14), we have

$$\int_{\mathscr{D}} \sigma(u_\varepsilon - u) \cdot (\nabla v)^s = (1 - \gamma)\int_{B_\varepsilon} \sigma(u_\varepsilon) \cdot (\nabla v)^s . \tag{5.34}$$

On the other hand, by setting $\eta = u_\varepsilon - u$ as test function in the adjoint equation (5.14), there is

$$\int_{\mathscr{D}} \sigma(v) \cdot (\nabla(u_\varepsilon - u))^s = -\int_{\Gamma_N} g \cdot (u_\varepsilon - u) . \tag{5.35}$$

Since the bilinear forms on the left-hand side of the above two last equations are symmetric, then we obtain the following important equality:

$$\int_{\Gamma_N} g \cdot (u_\varepsilon - u) = -(1 - \gamma) \int_{B_\varepsilon} \sigma(u_\varepsilon) \cdot (\nabla v)^s . \tag{5.36}$$

After comparing the above result with (5.30), we conclude that

$$\mathscr{J}_\varepsilon(u_\varepsilon) - \mathscr{J}(u) = -(1 - \gamma) \int_{B_\varepsilon} \sigma(u_\varepsilon) \cdot (\nabla v)^s . \tag{5.37}$$

Therefore, thanks to the adjoint state v solution of (5.14), the variation of the shape functional can, in fact, be written in terms of an integral concentrated in the ball B_ε. Before proceeding, let us sum and subtract the term

$$- (1 - \gamma) \int_{B_\varepsilon} \sigma(u) \cdot (\nabla v)^s \tag{5.38}$$

from (5.37) to obtain

$$\mathscr{J}_\varepsilon(u_\varepsilon) - \mathscr{J}(u) = -(1 - \gamma) \int_{B_\varepsilon} \sigma(u) \cdot (\nabla v)^s + \mathscr{I}(\varepsilon) . \tag{5.39}$$

The integral $\mathscr{I}(\varepsilon)$ is defined as

$$\mathscr{I}(\varepsilon) = -(1 - \gamma) \int_{B_\varepsilon} \sigma(u_\varepsilon - u) \cdot (\nabla v)^s , \tag{5.40}$$

which can be bounded as follows:

$$|\mathscr{I}(\varepsilon)| \leq C_1 \|\nabla v\|_{L^2(B_\varepsilon)} \|\sigma(u_\varepsilon - u)\|_{L^2(B_\varepsilon)}$$
$$\leq C_2 \varepsilon \|u_\varepsilon - u\|_{H^1(\Omega)} \leq C_3 \varepsilon^2 = O(\varepsilon^2) , \tag{5.41}$$

where we have used Lemma 5.1, together with the interior elliptic regularity of function u. According to Lemma 5.1, a leading term of order $O(\varepsilon^2)$ is expected. On the other hand, the above estimate cannot be improved, so that there is a nontrivial term of order $O(\varepsilon^2)$ hidden in (5.40). In the next section we will show how to extract such a leading term of order $O(\varepsilon^2)$ from (5.40).

5.1.3 Asymptotic Analysis of the Solution

The variation of the tracking-type shape functional has been written exclusively in terms of an integral concentrated in the ball B_ε, as shown through (5.37). In order to obtain the associated topological asymptotic expansion in the form of (1.2), we need to know the asymptotic behavior of the solution u_ε with respect to ε in the neighborhood of the ball B_ε. In particular, once knowing explicitly such a behavior, function $f(\varepsilon)$ can be identified, which allows for evaluating the limit $\varepsilon \to 0$ in (1.4), leading to the final formula for the topological derivative $\mathscr{T}$ of the shape functional ψ. Therefore, the basic idea consists in expanding u_ε asymptotically with respect to the small parameter ε. In this section, we obtain the asymptotic expansion of the solution u_ε associated with the transmission condition on the boundary ∂B_ε of the inclusion. We start by proposing an *ansatz* for u_ε in the form [58]

$$u_\varepsilon(x) = u(x) + w_\varepsilon(x) + \widetilde{u}_\varepsilon(x) . \tag{5.42}$$

After applying the operator $\sigma_\varepsilon = \gamma_\varepsilon \sigma$, we have

$$\sigma_\varepsilon(u_\varepsilon(x)) = \sigma_\varepsilon(u(x)) + \sigma_\varepsilon(w_\varepsilon(x)) + \sigma_\varepsilon(\widetilde{u}_\varepsilon(x))$$
$$= \gamma_\varepsilon \sigma(u(\widehat{x})) + \gamma_\varepsilon(\sigma(u(x)) - \sigma(u(\widehat{x})) + \sigma_\varepsilon(w_\varepsilon(x)) + \sigma_\varepsilon(\widetilde{u}_\varepsilon(x). \tag{5.43}$$

On the boundary of the inclusion ∂B_ε there is

$$[\![\sigma_\varepsilon(u_\varepsilon)]\!]n = 0 \quad \Rightarrow \quad (\sigma(u_\varepsilon)|_{\mathscr{D}\backslash\overline{B_\varepsilon}} - \gamma\sigma(u_\varepsilon)|_{B_\varepsilon})n = 0 , \tag{5.44}$$

so that the above expansion evaluated on ∂B_ε yields

$$(1 - \gamma)\sigma(u(\widehat{x}))n + (1 - \gamma)(\sigma(u(x)) - \sigma(u(\widehat{x}))n$$
$$+ [\![\sigma_\varepsilon(w_\varepsilon(x))]\!]n + [\![\sigma_\varepsilon(\widetilde{u}_\varepsilon(x))]\!]n = 0 , \tag{5.45}$$

which allows for choosing the jump $[\![\sigma_\varepsilon(w_\varepsilon(x))]\!]n$ on ∂B_ε as

$$[\![\sigma_\varepsilon(w_\varepsilon(x))]\!]n = -(1 - \gamma)\sigma(u(\widehat{x}))n \quad \text{on} \quad \partial B_\varepsilon . \tag{5.46}$$

Now, the following *exterior problem* is formally defined with $\varepsilon \to 0$: Find $\sigma_\varepsilon(w_\varepsilon)$, such that

$$\begin{cases} \operatorname{div} \sigma_\varepsilon(w_\varepsilon) = 0 \text{ in } \mathbb{R}^2 , \\ \sigma_\varepsilon(w_\varepsilon) \to 0 \text{ at } \infty , \\ [\![\sigma_\varepsilon(w_\varepsilon)]\!]n = \widehat{v} \text{ on } \partial B_\varepsilon , \end{cases} \tag{5.47}$$

Fig. 5.3 Polar coordinate
system (r, θ) centered at the
point $\widehat{x}$

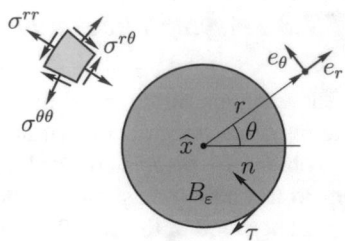

where $\widehat{v} = -(1-\gamma)\sigma(u(\widehat{x}))n$. The above boundary value problem admits an *explicit
solution* (see, for instance, the book by Little [62]), which can be written in a polar
coordinate system (r, θ) with center at $\widehat{x}$ (see Fig. 5.3) as follows:

- For $r \geq \varepsilon$ (outside the inclusion)

$$\sigma_\varepsilon^{rr}(w_\varepsilon(r, \theta)) = -\varphi_1 \left(\frac{1-\gamma}{1+\gamma a_1} \frac{\varepsilon^2}{r^2} \right)$$

$$-\varphi_2 \left(4\frac{1-\gamma}{1+\gamma a_2} \frac{\varepsilon^2}{r^2} + 3\frac{1-\gamma}{1+\gamma a_2} \frac{\varepsilon^4}{r^4} \right) \cos 2\theta , \qquad (5.48)$$

$$\sigma_\varepsilon^{\theta\theta}(w_\varepsilon(r, \theta)) = \varphi_1 \left(\frac{1-\gamma}{1+\gamma a_1} \frac{\varepsilon^2}{r^2} \right) - \varphi_2 \left(3\frac{1-\gamma}{1+\gamma a_2} \frac{\varepsilon^4}{r^4} \right) \cos 2\theta , \qquad (5.49)$$

$$\sigma_\varepsilon^{r\theta}(w_\varepsilon(r, \theta)) = -\varphi_2 \left(2\frac{1-\gamma}{1+\gamma a_2} \frac{\varepsilon^2}{r^2} - 3\frac{1-\gamma}{1+\gamma a_2} \frac{\varepsilon^4}{r^4} \right) \sin 2\theta . \qquad (5.50)$$

- For $0 < r < \varepsilon$ (inside the inclusion)

$$\sigma_\varepsilon^{rr}(w_\varepsilon(r, \theta)) = \varphi_1 \left(a_1\gamma \frac{1-\gamma}{1+\gamma a_1} \right) + \varphi_2 \left(a_2\gamma \frac{1-\gamma}{1+\gamma a_2} \right) \cos 2\theta , \qquad (5.51)$$

$$\sigma_\varepsilon^{\theta\theta}(w_\varepsilon(r, \theta)) = \varphi_1 \left(a_1\gamma \frac{1-\gamma}{1+\gamma a_1} \right) - \varphi_2 \left(a_2\gamma \frac{1-\gamma}{1+\gamma a_2} \right) \cos 2\theta , \qquad (5.52)$$

$$\sigma_\varepsilon^{r\theta}(w_\varepsilon(r, \theta)) = -\varphi_2 \left(a_2\gamma \frac{1-\gamma}{1+\gamma a_2} \right) \sin 2\theta . \qquad (5.53)$$

Some terms in the above formulae require explanations. The coefficients φ_1 and φ_2
are given by

$$\varphi_1 = \frac{1}{2}(\sigma_1(u(\widehat{x})) + \sigma_2(u(\widehat{x}))) , \quad \varphi_2 = \frac{1}{2}(\sigma_1(u(\widehat{x})) - \sigma_2(u(\widehat{x}))) , \qquad (5.54)$$

where $\sigma_1(u(\widehat{x}))$ and $\sigma_2(u(\widehat{x}))$ are the eigenvalues of tensor $\sigma(u(\widehat{x}))$, which can be
expressed as (see Appendix A, identity (A.52))

$$\sigma_{1,2}(u(\widehat{x})) = \frac{1}{2} \left(\operatorname{tr} \sigma(u(\widehat{x})) \pm \sqrt{2\sigma^D(u(\widehat{x})) \cdot \sigma^D(u(\widehat{x}))} \right) , \qquad (5.55)$$

with $\sigma^D(u(\widehat{x}))$ standing for the deviatoric part of the stress tensor $\sigma(u(\widehat{x}))$, namely

$$\sigma^D(u(\widehat{x})) = \sigma(u(\widehat{x})) - \frac{1}{2}\mathrm{tr}\,\sigma(u(\widehat{x}))\mathrm{I}\,. \tag{5.56}$$

In addition, the constants a_1 and a_2 are given by

$$a_1 = \frac{\mu + \lambda}{\mu} \quad \text{e} \quad a_2 = \frac{3\mu + \lambda}{\mu + \lambda}\,. \tag{5.57}$$

Finally, $\sigma_\varepsilon^{rr}(u_\varepsilon)$, $\sigma_\varepsilon^{\theta\theta}(u_\varepsilon)$, and $\sigma_\varepsilon^{r\theta}(u_\varepsilon)$ are the components of tensor $\sigma_\varepsilon(u_\varepsilon)$ in the polar coordinate system, namely $\sigma_\varepsilon^{rr}(u_\varepsilon) = e^r \cdot \sigma_\varepsilon(u_\varepsilon)e^r$, $\sigma_\varepsilon^{\theta\theta}(u_\varepsilon) = e^\theta \cdot \sigma_\varepsilon(u_\varepsilon)e^\theta$, and $\sigma_\varepsilon^{r\theta}(u_\varepsilon) = \sigma_\varepsilon^{\theta r}(u_\varepsilon) = e^r \cdot \sigma_\varepsilon(u_\varepsilon)e^\theta$, with e^r and e^θ used to denote the canonical basis associated with the polar coordinate system (r, θ), such that, $\|e^r\| = \|e^\theta\| = 1$ and $e^r \cdot e^\theta = 0$. See Appendix A.

Remark 5.4 (Eshelby's Theorem) According to (5.51)–(5.53), we observe that the stress tensor field associated with the solution of the exterior problem (5.47) is uniform inside the inclusion $B_\varepsilon(\widehat{x})$. It means that the stress acting in the inclusion embedded in the whole two-dimensional space $\mathbb{R}^2$ can be written in the following compact form:

$$\sigma_\varepsilon(w_\varepsilon(x))|_{B_\varepsilon(\widehat{x})} = \gamma\mathbb{T}\sigma(u(\widehat{x}))\,, \tag{5.58}$$

where $\mathbb{T}$ is a fourth order uniform (constant) tensor given by

$$\mathbb{T} = \frac{1}{2}\frac{1-\gamma}{1+\gamma a_2}\left(2a_2\mathbb{I} + \frac{a_1 - a_2}{1 + \gamma a_1}\mathrm{I} \otimes \mathrm{I}\right)\,. \tag{5.59}$$

Therefore, the above result fits the famous Eshelby's problem. This problem, formulated by Eshelby in 1957 [38] and 1959 [39], represents one of the major advances in the continuum mechanics theory of the twentieth century [56]. It plays a central role in the theory of elasticity involving the determination of effective elastic properties of materials with multiple inhomogeneities. For more details, see the book by Mura [70], for instance. The Eshelby's problem, also referred to as the Eshelby's theorem, is also related to the Polarization tensor in asymptotic analysis of the strain energy with respect to singular domain perturbations [71]. In fact, tensor $\mathbb{T}$ represents one term contribution to the Polarization tensor coming from the solution to the exterior problem (5.47). In the next section we will apply the Eshelby's theorem to the derivation of the polarization tensor and to the topological derivative evaluation as well. Concerning applications of the Eshelby's theorem to the problem of optimal patch in elasticity, see [61, 72].

Now, we can construct the remainder $\tilde{u}_\varepsilon$ from (5.42) in such a way that it compensates for the discrepancies produced by the higher order terms in ε as well as by the boundary layer w_ε on the exterior boundary $\partial\mathscr{D}$. It means that the *remainder* $\tilde{u}_\varepsilon$ has to be the solution of the following boundary value problem: Find $\tilde{u}_\varepsilon$, such that

$$\begin{cases} \operatorname{div}\sigma_\varepsilon(\tilde{u}_\varepsilon) = 0 & \text{in } \mathscr{D}, \\ \sigma_\varepsilon(\tilde{u}_\varepsilon) = \gamma_\varepsilon\rho\mathbb{C}(\nabla\tilde{u}_\varepsilon)^s & \\ \tilde{u}_\varepsilon = f_\varepsilon & \text{on } \Gamma_D, \\ \sigma(\tilde{u}_\varepsilon)n = g_\varepsilon & \text{on } \Gamma_N, \\ \left.\begin{array}{l}[\![\tilde{u}_\varepsilon]\!] = 0 \\ [\![\sigma_\varepsilon(\tilde{u}_\varepsilon)]\!]n = h_\varepsilon\end{array}\right\} & \text{on } \partial B_\varepsilon, \end{cases} \tag{5.60}$$

where $f_\varepsilon = -w_\varepsilon|_{\Gamma_D}$, $g_\varepsilon = -\sigma(w_\varepsilon)n|_{\Gamma_N}$, and $h_\varepsilon = \tilde{\sigma}n$, with the second order tensor field $\tilde{\sigma}(x) = -(1-\gamma)[\sigma(u(x)) - \sigma(u(\hat{x}))]$. From the above boundary value problem, it is possible to prove that the remainder $\tilde{u}_\varepsilon$ enjoys an estimate of the form $\tilde{u}_\varepsilon \approx O(\varepsilon^2)$ in an appropriated norm. In fact, before continuing, let us state the following important result:

Lemma 5.2 *Let $\tilde{u}_\varepsilon$ be the solution of (5.60) or equivalently solution of the following variational problem:*

$$\tilde{u}_\varepsilon \in \mathscr{U}_\varepsilon : \int_{\mathscr{D}}\sigma_\varepsilon(\tilde{u}_\varepsilon)\cdot(\nabla\eta)^s = \int_{\Gamma_N}g_\varepsilon\cdot\eta + \int_{\partial B_\varepsilon}h_\varepsilon\cdot\eta \quad \forall\eta\in\mathscr{V}_\varepsilon, \tag{5.61}$$

with $\sigma_\varepsilon(\tilde{u}_\varepsilon) = \gamma_\varepsilon\mathbb{C}(\nabla\tilde{u}_\varepsilon)^s$, where the set $\mathscr{U}_\varepsilon$ and the space $\mathscr{V}_\varepsilon$ are defined respectively as

$$\mathscr{U}_\varepsilon := \{\varphi\in H^1(\mathscr{D}) : \varphi|_{\Gamma_D} = f_\varepsilon\},$$

$$\mathscr{V}_\varepsilon := \{\varphi\in H^1(\mathscr{D}) : \varphi|_{\Gamma_D} = 0\}.$$

Then, we have that the following estimate for the remainder $\tilde{u}_\varepsilon$ holds true:

$$\|\tilde{u}_\varepsilon\|_{H^1(\mathscr{D})} \leq C\varepsilon^2, \tag{5.62}$$

with constant C independent of the small parameter ε.

Proof From the definition of function $h_\varepsilon = \tilde{\sigma}n$, with n used to denote the unit normal vector field on ∂B_ε pointing toward to the center of the inclusion, we have

$$\int_{\partial B_\varepsilon}h_\varepsilon\cdot\eta = \int_{\partial B_\varepsilon}\tilde{\sigma}n\cdot\eta = -\int_{B_\varepsilon}\operatorname{div}(\tilde{\sigma}\eta) = -\int_{B_\varepsilon}\operatorname{div}(\tilde{\sigma})\cdot\eta - \int_{B_\varepsilon}\tilde{\sigma}\cdot(\nabla\eta)^s$$

$$= (1-\gamma)\int_{B_\varepsilon}\operatorname{div}(\sigma(u))\cdot\eta + (1-\gamma)\int_{B_\varepsilon}[\sigma(u) - \sigma(u(\hat{x}))]\cdot(\nabla\eta)^s, \tag{5.63}$$

where we have taken into account that $\widetilde{\sigma}(x) = -(1-\gamma)[\sigma(u(x)) - \sigma(u(\widehat{x}))]$. From this last result, the variational form (5.64) can be rewritten as follows:

$$\widetilde{u}_\varepsilon \in \mathcal{U}_\varepsilon : \int_\mathcal{D} \sigma_\varepsilon(\widetilde{u}_\varepsilon) \cdot (\nabla\eta)^s = \int_{\Gamma_N} g_\varepsilon \cdot \eta - \int_{B_\varepsilon} \widetilde{\sigma} \cdot (\nabla\eta)^s \quad \forall \eta \in \mathcal{V}_\varepsilon, \quad (5.64)$$

since $\text{div}\,(\sigma(u)) = 0$. By taking $\eta = \widetilde{u}_\varepsilon - \varphi_\varepsilon$ as test function in (5.64), where $\varphi_\varepsilon \in \mathcal{U}_\varepsilon$ is the lifting of the Dirichlet boundary data f_ε on Γ_D, we have

$$\int_\mathcal{D} \sigma_\varepsilon(\widetilde{u}_\varepsilon) \cdot (\nabla\widetilde{u}_\varepsilon)^s = \int_{\Gamma_D} f_\varepsilon \cdot \sigma(\widetilde{u}_\varepsilon)n + \int_{\Gamma_N} g_\varepsilon \cdot \widetilde{u}_\varepsilon - \int_{B_\varepsilon} \widetilde{\sigma} \cdot (\nabla\widetilde{u}_\varepsilon)^s. \quad (5.65)$$

From the *Cauchy–Schwarz inequality* and the *trace theorem* there are

$$\left|\int_{\Gamma_D} f_\varepsilon \cdot \sigma(\widetilde{u}_\varepsilon)n\right| \leq \|f_\varepsilon\|_{H^{1/2}(\Gamma_D)} \|\sigma(\widetilde{u}_\varepsilon)n\|_{H^{-1/2}(\Gamma_D)}$$

$$\leq C_1\varepsilon^2 \|\nabla\widetilde{u}_\varepsilon\|_{L^2(\mathcal{D})} \leq C_2\varepsilon^2 \|\widetilde{u}_\varepsilon\|_{H^1(\mathcal{D})}, \quad (5.66)$$

and

$$\left|\int_{\Gamma_N} g_\varepsilon \cdot \widetilde{u}_\varepsilon\right| \leq \|g_\varepsilon\|_{H^{-1/2}(\Gamma_N)} \|\widetilde{u}_\varepsilon\|_{H^{1/2}(\Gamma_N)} \leq C_3\varepsilon^2 \|\widetilde{u}_\varepsilon\|_{H^1(\mathcal{D})}, \quad (5.67)$$

where we have used the fact that f_ε and g_ε have order $O(\varepsilon^2)$ on the exterior boundary $\partial\mathcal{D}$. By taking into account the definition $\widetilde{\sigma}(x) = -(1-\gamma)[\sigma(u(x)) - \sigma(u(\widehat{x}))]$, there is

$$\left|\int_{B_\varepsilon} \widetilde{\sigma} \cdot (\nabla\widetilde{u}_\varepsilon)^s\right| \leq \|\widetilde{\sigma}\|_{L^2(B_\varepsilon)} \|\nabla\widetilde{u}_\varepsilon\|_{L^2(B_\varepsilon)}$$

$$\leq C_4 \|\sigma(u) - \sigma(u(\widehat{x}))\|_{L^2(B_\varepsilon)} \|\nabla\widetilde{u}_\varepsilon\|_{L^2(B_\varepsilon)}$$

$$\leq C_5 \|x - \widehat{x}\|_{L^2(B_\varepsilon)} \|\nabla\widetilde{u}_\varepsilon\|_{L^2(B_\varepsilon)} \leq C_6\varepsilon^2 \|\widetilde{u}_\varepsilon\|_{H^1(\mathcal{D})}, \quad (5.68)$$

where we have used again the Cauchy–Schwarz inequality together with the interior elliptic regularity of function u. From these results, we obtain

$$\int_\mathcal{D} \sigma_\varepsilon(\widetilde{u}_\varepsilon) \cdot (\nabla\widetilde{u}_\varepsilon)^s \leq C_7\varepsilon^2 \|\widetilde{u}_\varepsilon\|_{H^1(\mathcal{D})}. \quad (5.69)$$

Finally, from the *coercivity* of the bilinear form on the left-hand side of the above inequality, namely

$$c\|\widetilde{u}_\varepsilon\|^2_{H^1(\mathcal{D})} \leq \int_\mathcal{D} \sigma_\varepsilon(\widetilde{u}_\varepsilon) \cdot (\nabla\widetilde{u}_\varepsilon)^s, \quad (5.70)$$

we obtain the result with $C = C_7/c$. $\square$

5.1.4 Topological Derivative Evaluation

From the above elements, the integral (5.40) can be evaluated explicitly, which allows for collecting the terms in power of ε. Thus, it is possible to identify the function $f(\varepsilon)$ in (1.2) and compute the limit passage $\varepsilon \to 0$, leading to the final formula for the associated topological derivative. In particular, the integral (5.40) can be rewritten as

$$\mathscr{I}(\varepsilon) = -\frac{1-\gamma}{\gamma} \int_{B_\varepsilon} \sigma_\varepsilon (u_\varepsilon - u) \cdot (\nabla v)^s , \qquad (5.71)$$

where we have used the definition for the contrast given by (5.3). After replacing the expansion (5.42) into the above equation we obtain

$$\mathscr{I}(\varepsilon) = -\frac{1-\gamma}{\gamma} \int_{B_\varepsilon} \sigma_\varepsilon (w_\varepsilon + \widetilde{u}_\varepsilon) \cdot (\nabla v)^s$$

$$= -\frac{1-\gamma}{\gamma} \int_{B_\varepsilon} \sigma_\varepsilon (w_\varepsilon) \cdot (\nabla v)^s + \mathscr{E}_1(\varepsilon) . \qquad (5.72)$$

The remainder $\mathscr{E}_1(\varepsilon)$ is defined as

$$\mathscr{E}_1(\varepsilon) = -\frac{1-\gamma}{\gamma} \int_{B_\varepsilon} \sigma_\varepsilon (\widetilde{u}_\varepsilon) \cdot (\nabla v)^s . \qquad (5.73)$$

The Cauchy–Schwarz inequality together with the interior elliptic regularity of function u yield

$$|\mathscr{E}_1(\varepsilon)| \le C_1 \|\nabla v\|_{L^2(B_\varepsilon)} \|\sigma_\varepsilon (\widetilde{u}_\varepsilon)\|_{L^2(B_\varepsilon)} \le C_2 \varepsilon \|\nabla \widetilde{u}_\varepsilon\|_{L^2(\Omega)} . \qquad (5.74)$$

From Lemma 5.2, we have

$$|\mathscr{E}_1(\varepsilon)| \le C_3 \varepsilon \|\widetilde{u}_\varepsilon\|_{H^1(\Omega)} \le C_4 \varepsilon^3 = O(\varepsilon^3) . \qquad (5.75)$$

Now, let us comeback to the *expansion* (5.39), which can be written as

$$\mathscr{J}_\varepsilon(u_\varepsilon) - \mathscr{J}(u) = -(1-\gamma) \int_{B_\varepsilon} \sigma(u) \cdot (\nabla v)^s - \frac{1-\gamma}{\gamma} \int_{B_\varepsilon} \sigma_\varepsilon(w_\varepsilon) \cdot (\nabla v)^s + \mathscr{E}_1(\varepsilon)$$

$$= -(1-\gamma) \int_{B_\varepsilon} (\mathbb{I} + \mathbb{T}) \sigma(u(\widehat{x})) \cdot (\nabla v(\widehat{x}))^s + \sum_{i=1}^{3} \mathscr{E}_i(\varepsilon)$$

$$= \pi \varepsilon^2 \mathbb{P}_\gamma \sigma(u(\widehat{x})) \cdot (\nabla v(\widehat{x}))^s + \sum_{i=1}^{3} \mathscr{E}_i(\varepsilon) , \qquad (5.76)$$

with $\mathbb{P}_\gamma = -(1-\gamma)(\mathbb{I}+\mathbb{T})$, where we have used the explicit solution for $\sigma_\varepsilon(w_\varepsilon)|_{B_\varepsilon}$ given by (5.58). The remainders $\mathscr{E}_2(\varepsilon)$ and $\mathscr{E}_3(\varepsilon)$ are respectively defined as

$$\mathscr{E}_2(\varepsilon) = -(1-\gamma)\int_{B_\varepsilon} (\sigma(u) \cdot (\nabla v)^s - \sigma(u(\widehat{x})) \cdot (\nabla v(\widehat{x}))^s) , \qquad (5.77)$$

$$\mathscr{E}_3(\varepsilon) = -\frac{1-\gamma}{\gamma}\int_{B_\varepsilon} \sigma_\varepsilon(w_\varepsilon) \cdot ((\nabla v)^s - (\nabla v(\widehat{x}))^s) , \qquad (5.78)$$

which can be trivially bounded as follows:

$$|\mathscr{E}_2(\varepsilon)| \le C_1\varepsilon^3 = O(\varepsilon^3) , \qquad (5.79)$$

$$|\mathscr{E}_3(\varepsilon)| \le C_2\varepsilon^3 = O(\varepsilon^3) , \qquad (5.80)$$

where we have used the interior elliptic regularity of function u and the explicit solution (5.58). According to the estimates (5.75) and (5.79), the remainders $\mathscr{E}_1(\varepsilon)$, $\mathscr{E}_2(\varepsilon)$, and $\mathscr{E}_3(\varepsilon)$ are of order $o(\varepsilon^2)$. Therefore, from the expansion (5.76) we promptly identify function $f(\varepsilon) = \pi\varepsilon^2$ and thus the final formula for the *topological derivative* as [9, 45]

$$\mathscr{T}(\widehat{x}) = \mathbb{P}_\gamma\sigma(u(\widehat{x})) \cdot (\nabla v(\widehat{x}))^s \quad \forall \widehat{x} \in \Omega , \qquad (5.81)$$

where the *polarization tensor* $\mathbb{P}_\gamma$ is given by the following fourth order isotropic tensor:

$$\mathbb{P}_\gamma = -\frac{1-\gamma}{1+\gamma a_2}\left((1+a_2)\mathbb{I} + \frac{1}{2}(a_1-a_2)\frac{1-\gamma}{1+\gamma a_1}I\otimes I\right) , \qquad (5.82)$$

with the parameters a_1 and a_2 given by (5.57).

Remark 5.5 Note that the polarization tensor defined through (5.82) is isotropic because we are dealing with circular inclusions. For the polarization tensor regarding arbitrary-shaped inclusions, the reader may refer to the book by Ammari and Kang [5], for instance.

Remark 5.6 Formally, we can evaluate the limits $\gamma \to 0$ and $\gamma \to \infty$ in (5.82). For $\gamma \to 0$, the inclusion becomes a void and the transmission condition on the interface of the inclusion degenerates itself to the homogeneous Neumann boundary condition on the boundary of the resulting hole $B_\varepsilon(\widehat{x})$. Thus, in this particular case the *polarization tensor* is given by

$$\begin{aligned}\mathbb{P}_0 &= -(1+a_2)\mathbb{I} - \frac{a_1-a_2}{2}I\otimes I \\ &= -\frac{2\mu+\lambda}{\mu+\lambda}\left(2\mathbb{I} - \frac{\mu-\lambda}{2\mu}I\otimes I\right) . \end{aligned} \qquad (5.83)$$

In addition, for $\gamma \rightarrow \infty$, the elastic inclusion becomes a rigid one and the polarization tensor is given by

$$
\begin{aligned}
\mathbb{P}_\infty &= \frac{1 + a_2}{a_2} \mathbb{I} - \frac{a_1 - a_2}{2a_1 a_2} I \otimes I \\
&= \frac{2\mu + \lambda}{3\mu + \lambda} \left(2\mathbb{I} + \frac{\mu - \lambda}{2(\mu + \lambda)} I \otimes I \right) .
\end{aligned} \tag{5.84}
$$

The rigorous mathematical justification for these limit cases can be found in [7], for instance.

5.2 Topology Design Algorithm

In this section a topology optimization algorithm based on the topological derivative combined with a level-set domain representation method is presented. It has been proposed by Amstutz and Andrä [11] and consists basically in achieving a local optimality condition for the minimization problem (5.2), given in terms of the topological derivative and a level-set function. In particular, the domain $\Omega \subset \mathscr{D}$ and the complement $\mathscr{D} \setminus \Omega$ are characterized by a *level-set function* Ψ:

$$
\Omega = \{ x \in \mathscr{D} : \Psi(x) < 0 \} \quad \text{and} \quad \mathscr{D} \setminus \Omega = \{ x \in \mathscr{D} : \Psi(x) > 0 \}, \tag{5.85}
$$

where Ψ vanishes on the interface between Ω and $\mathscr{D} \setminus \Omega$. A local sufficient *optimality condition* for Problem (5.2), under a class of domain perturbations given by ball-shaped inclusions denoted by $B_\varepsilon(x)$, can be stated as [10]

$$
\mathscr{T}(x) > 0 \quad \forall x \in \mathscr{D}, \tag{5.86}
$$

where $\mathscr{T}(x)$ is the topological derivative of the shape functional $J(\Omega)$ at $x \in \mathscr{D}$ and $B_\varepsilon(x)$ is a ball of radius ε centered at $x \in \mathscr{D}$, as shown in Fig. 5.4. Therefore, let us define the quantity

$$
g(x) := \begin{cases} -\mathscr{T}(x) \text{ if } \Psi(x) < 0, \\ +\mathscr{T}(x) \text{ if } \Psi(x) > 0, \end{cases} \tag{5.87}
$$

which allows rewriting the condition (5.86) in the following equivalent form:

$$
\begin{cases} g(x) < 0 \text{ if } \Psi(x) < 0, \\ g(x) > 0 \text{ if } \Psi(x) > 0. \end{cases} \tag{5.88}
$$

We observe that (5.88) is satisfied, where the quantity g coincides with the level-set function Ψ up to a strictly positive factor, namely $\exists \, \tau > 0 : g = \tau \Psi$, or equivalently

Fig. 5.4 Nucleation of a ball-shaped inclusion $B_\varepsilon(x)$

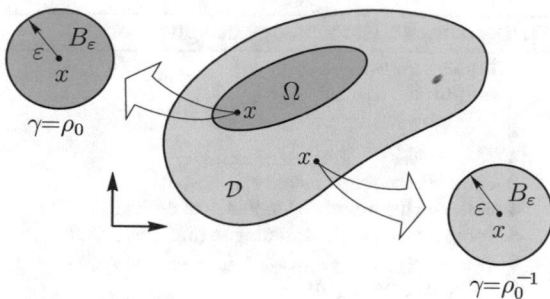

$$\theta := \arccos\left[\frac{\langle g, \Psi \rangle_{L^2(\mathscr{D})}}{\|g\|_{L^2(\mathscr{D})}\|\Psi\|_{L^2(\mathscr{D})}}\right] = 0, \tag{5.89}$$

which will be used as the optimality condition in the topology design algorithm, where θ is the angle in $L^2(\mathscr{D})$ between the functions g and Ψ.

Let us now explain the algorithm. We start by choosing an initial level-set function Ψ_0. In a generic iteration n, we compute the function g_n associated with the level-set function Ψ_n. Thus, the new level-set function Ψ_{n+1} is updated according to the following linear combination between the functions g_n and Ψ_n:

$$\Psi_0 : \|\Psi_0\|_{L^2(\mathscr{D})} = 1,$$
$$\Psi_{n+1} = \frac{1}{\sin\theta_n}\left[\sin((1-k)\theta_n)\Psi_n + \sin(k\theta_n)\frac{g_n}{\|g_n\|_{L^2(\mathscr{D})}}\right] \quad \forall n \in \mathbb{N}, \tag{5.90}$$

where θ_n is the angle between g_n and Ψ_n, and k is a step size determined by a line-search performed in order to decrease the value of the objective function $J(\Omega_n)$, with Ω_n used to denote the domain associated with Ψ_n. The process ends when the condition $\theta_n \leq \epsilon_\theta$ is satisfied at some iteration, where ϵ_θ is a given small numerical tolerance. Since we have chosen $\Psi_0 : \|\Psi_0\|_{L^2(\mathscr{D})} = 1$, by construction $\Psi_{n+1} : \|\Psi_{n+1}\|_{L^2(\mathscr{D})} = 1 \ \forall n \in \mathbb{N}$. If at some iteration n the line-search step size k is found to be smaller, then a given numerical tolerance $\epsilon_k > 0$ and the optimality condition is not satisfied, namely $\theta_n > \epsilon_\theta$, then a mesh refinement of the hold-all domain $\mathscr{D}$ is carried out and the iterative process is continued. The resulting *topology design algorithm* is summarized in pseudo-code format in Algorithm 1. For further applications of this algorithm, see for instance [4, 14, 17, 49, 64, 85, 92].

In the context of topological-derivative-based topology optimization methods, the algorithms available in the literature usually combine topological derivatives with shape derivatives or level-set methods [1, 25, 36], leading to a two-stage shape/topology optimization procedure. More precisely, new holes are nucleated according to the topological derivative, while standard tools in shape optimization are used to move the new boundaries. In contrast, Algorithm 1 is based on the optimality condition (5.86) written in terms of the topological derivative and a level-set function, leading to a very simple and quite efficient one-stage algorithm driven

---

Algorithm 1: The topology design algorithm

---
input : $\mathcal{D}$, Ψ_0, ϵ_k, ϵ_θ;
output: the optimal topology $\Omega^\star$;

1 $n \leftarrow 0$;
2 $\Omega_n \leftarrow \Psi_n$;
3 compute the shape functional $J(\Omega_n)$;
4 compute the associated topological derivative $\mathcal{T}(x)$;
5 compute g_n and θ_n according to (5.87) and (5.89);
6 $\Psi_{\text{old}} \leftarrow \Psi_n$; $J_{\text{old}} \leftarrow J(\Omega_n)$; $J_{\text{new}} \leftarrow 1 + J_{\text{old}}$; $k \leftarrow 1$;
7 **while** $J_{\text{new}} > J_{\text{old}}$ **do**
8 | compute Ψ_{new} according to (5.90);
9 | $\Psi_n \leftarrow \Psi_{\text{new}}$;
10 | execute lines 2 and 3;
11 | $J_{\text{new}} \leftarrow J(\Omega_n)$;
12 | $k \leftarrow k/2$;
13 **end while**
14 **if** $k < \epsilon_k$ **then**
15 | try a mesh refinement;
16 | $\Psi_{n+1} \leftarrow \Psi_n$; $n \leftarrow n + 1$;
17 | **go to** line 2;
18 **else if** $\theta_n > \epsilon_\theta$ **then**
19 | $\Psi_{n+1} \leftarrow \Psi_n$; $n \leftarrow n + 1$;
20 | **go to** line 2;
21 **else**
22 | **return** $\Omega^\star \leftarrow \Psi_n$;
23 | **stop;**
24 **end if**

---

by the topological derivative only. However, how to efficiently use the topological derivative in the context of topology optimization deserves further investigation [19]. See Sect. 5.4 for an account of some open problems.

5.3 Numerical Results

The topological derivative has been specifically designed to deal with shape and topology optimization problems [1, 23, 25, 47, 57, 60, 73, 74, 76–78, 93]. In contrast to traditional topology optimization methods, the topological derivative formulation does not require a material model concept based on intermediary densities, so that interpolation schemes are unnecessary. These features are crucial in a wide range of applications, since the limitations arising from material model procedures are here naturally avoided. In addition, topological derivative has the advantage of providing an analytical form for the topological sensitivity which allows to obtain the optimal design in a few iterations or even in just one shot. Therefore, the resulting topology optimization algorithms are remarkably efficient and of simple computational implementation, since it features only a minimal number of user-defined algorithmic parameters, as shown in Sect. 5.2, for instance. In this section, Algorithm 1 is applied in the context of compliance structural topology optimization and topology design of compliant mechanisms. In particular, the topology optimization problem we are dealing with consists in finding a subdomain $\Omega \subset \mathcal{D}$ that solves the following *minimization problem*:

$$\underset{\Omega \subset \mathcal{D}}{\text{Minimize}} \ \mathcal{F}_\Omega(u) = \mathcal{J}(u) + \beta |\Omega| , \tag{5.91}$$

where $\mathcal{J}(u)$ will be specified according to the application we are dealing with and $\beta > 0$ is a fixed multiplier used to impose a *volume constraint* in Ω of the form $|\Omega| \leq M$, with $M > 0$. In particular, by fixing different values of β we get different volume fractions at the end of the iterative process. For more sophisticated topological-derivative-based methods with volume constraint we refer the reader to [27], for instance. Since the last term in (5.91) represents the volume constraint, its associated topological derivative $\mathcal{T}_V(x)$ is trivially given by

$$\mathcal{T}_V(x) = \begin{cases} -1 \text{ if } x \in \Omega, \\ +1 \text{ if } x \in \mathcal{D} \setminus \Omega. \end{cases} \tag{5.92}$$

On the other hand, the first term in (5.91) depends on the state u solution of (5.6), so that the derivation of its topological derivative becomes much more involved, as presented in this chapter. Therefore, in this section we will adapt the obtained result (5.81) in such a way that it can be directly applied in the context of compliance structural topology optimization as well as in topology design of compliant mechanisms.

5.3.1 Structural Compliance Topology Optimization

Minimizing the structural flexibility under volume constraint is probably the most studied problem in the context of topology optimization. See the pioneering papers [20, 22] and also the book by Bendsøe [21], for instance. This classical problem is revisited here.

We start by setting $g = q$ in (5.14), which implies immediately that the adjoint state v, solution of (5.14), can be obtained as $v = -u$. See discussion in Remark 5.3. In this particular case, $\mathcal{J}(u)$ in (5.5) becomes the so-called compliance *shape functional*, namely

$$\mathcal{J}(u) = \int_{\Gamma_N} q \cdot u , \tag{5.93}$$

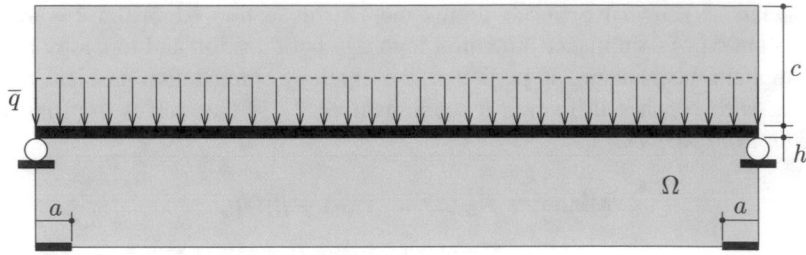

Fig. 5.5 Bridge design problem: initial guess and boundary conditions

where u is the solution of (5.6) and q is a given traction on Γ_N. By taking into account Remark 5.3 in result (5.81), the topological derivative of the compliance shape functional, denoted as $\mathscr{T}_C$, is given by

$$\mathscr{T}_C(x) = -\mathbb{P}_\gamma \sigma(u(x)) \cdot (\nabla u(x))^s \, , \qquad (5.94)$$

where $\mathbb{P}_\gamma$ is the polarization tensor defined through (5.82). Finally, the topological derivative of the shape functional $\mathscr{F}_\Omega(u)$ in (5.91) is obtained from the sum

$$\mathscr{T}(x) = \mathscr{T}_C(x) + \beta \mathscr{T}_V(x) \quad \forall x \in \mathscr{D} \, , \qquad (5.95)$$

where $\mathscr{T}_V(x)$ and $\mathscr{T}_C(x)$ are given by (5.92) and (5.94), respectively.

Let us now present a numerical example concerning the optimal *design of a bridge* structure borrowed from [75, Ch. 5, Sec. 5.2.5, p. 159]. The initial domain shown in Fig. 5.5 is represented by a rectangular panel of dimensions $180 \times 60 \, \text{m}^2$, which is clamped on the region $a = 9 \, \text{m}$ and submitted to a uniformly distributed traffic loading $q = 250 \times 10^3 \, \text{N/m}$. This load is applied on the dark strip of height $h = 3 \, \text{m}$, which is placed at a distance $c = 30 \, \text{m}$ from the top of the design domain. The dark strip will not be optimized. The Young modulus E and the Poisson ratio ν are set as $E = 210 \times 10^9 \, \text{N/m}^2$ and $\nu = 1/3$, respectively. The penalty parameter which appears in (5.91) is fixed to be $\beta = 10 \times 10^6$ and the contrast in (5.4) is set as $\rho_0 = 10^{-4}$. The topological derivative of the shape functional $\mathscr{F}_\Omega(u)$ obtained at the first iteration of the shape and topology optimization numerical procedure is shown in Fig. 5.6, where white to black levels mean smaller (negative) to higher (positive) values. This picture induces a level-set domain representation for the optimal shape, as proposed in [11]. See Algorithm 1. The resulting topology design obtained in the form of a well-known tied-arch bridge structure, which is acceptable from practical point of view, is shown in Fig. 5.7. Usually it is a local minimizer obtained numerically for the compliance minimization with volume constraint. Indeed, there is a lack of sufficient optimality conditions for such shape optimization problems [15]. The convergence curves for the angle θ_n and shape functional $J(\Omega_n)$ are shown in Fig. 5.8, where the picks come out from the mesh refinement procedure.

Fig. 5.6 Bridge design problem: topological derivative in the hold-all domain

Fig. 5.7 Bridge design problem: optimal domain [75]

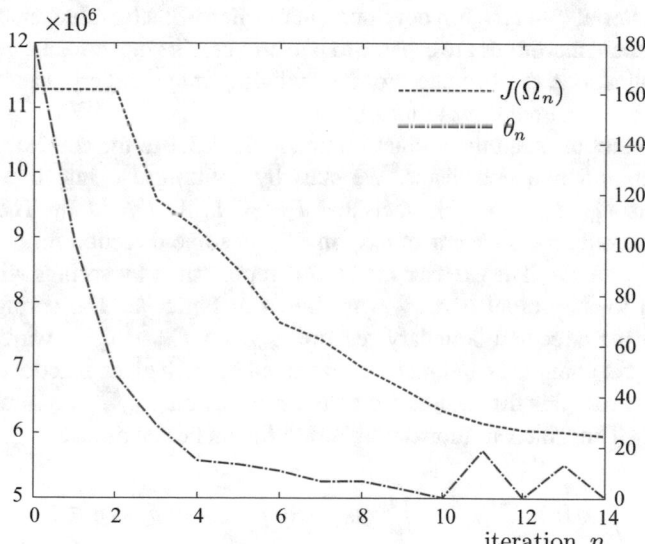

Fig. 5.8 Bridge design problem: convergence curves for the angle θ_n (dashed-dot red line) and shape functional $J(\Omega_n)$ (dashed blue line)

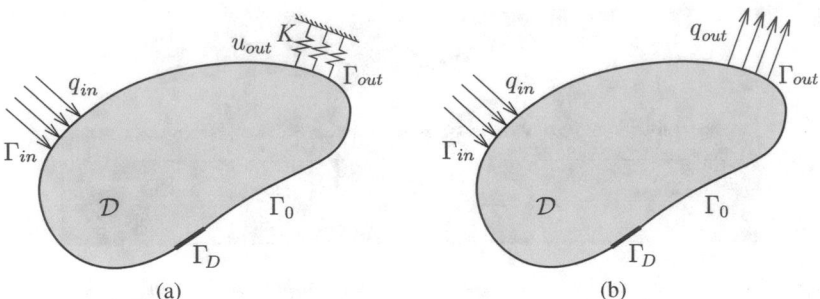

Fig. 5.9 Design of compliant mechanisms: problem setting. (**a**) Original model. (**b**) Surrogate model

5.3.2 Topology Design of Compliant Mechanisms

Compliant mechanisms are mechanical devices composed by one single peace that transforms simple inputs into complex movements by amplifying and changing their direction [2, 26, 29, 59, 63, 65, 69, 87]. Hence they are easy to fabricate and miniaturize and have no need for lubrication. Although these ideas are not new [26], compliant mechanisms have received considerable attention in recent years. This fact is due to manufacturing at a very small scale, the introduction of new advanced materials, and the fast development of Micro-Electro-Mechanical Systems [37]. Since such microtools are capable to perform precise movements, the spectrum of their applications has become broader including microsurgery, nanotechnology processing, cell manipulation, among others.

Therefore, let us adapt the problem stated in Sect. 5.1 to the context of topology design of compliant mechanisms. We start by splitting Γ_N into three mutually disjoint parts Γ_{in}, Γ_{out}, and Γ_0, such that $\Gamma_N = \Gamma_{in} \cup \Gamma_{out} \cup \Gamma_0$. The idea is to maximize the output displacement u_{out} on Γ_{out} in some direction for a given input excitation q_{in} on Γ_{in}. The exterior medium is represented by springs with stiffness K, attached to the output port Γ_{out}, as shown in Fig. 5.9a. The springs are then replaced by the expected boundary reaction q_{out} on Γ_{out}. In this way, the output displacement is going to be indirectly constrained by such given reaction. See sketch in Fig. 5.9b. From this discussion, we define $q = q_{in}$ on Γ_{in}, $q = q_{out}$ on Γ_{out}, and $q = 0$ on Γ_0. Thus, the variational problem (5.6) can be rewritten as

$$u \in \mathscr{V} : \int_{\Omega} \sigma(u) \cdot (\nabla \eta)^s = \int_{\Gamma_{in}} q_{in} \cdot \eta + \int_{\Gamma_{out}} q_{out} \cdot \eta \quad \forall \eta \in \mathscr{V} , \qquad (5.96)$$

with $\sigma(u) = \rho \mathbb{C}(\nabla u)^s$. In addition, we set $g = q_{in}$ on Γ_{in}, $g = \kappa q_{out}$ on Γ_{out}, and $g = 0$ on Γ_0, so that the *shape function* (5.5) becomes

$$\mathscr{J}(u) = \int_{\Gamma_{in}} q_{in} \cdot u + \kappa \int_{\Gamma_{out}} q_{out} \cdot u . \qquad (5.97)$$

Finally, the associated adjoint system (5.14) can be stated as

$$v \in \mathcal{V} : \int_{\Omega} \sigma(v) \cdot (\nabla \eta)^s = - \int_{\Gamma_{in}} q_{in} \cdot \eta - \kappa \int_{\Gamma_{out}} q_{out} \cdot \eta \in \mathcal{V} , \qquad (5.98)$$

with $\sigma(v) = \rho \mathbb{C}(\nabla v)^s$, where $\kappa > 0$ is a weight parameter. For more details concerning the adopted formulation, the reader may refer to [63], for instance. In this particular case, the topological derivative of the shape functional $\mathcal{F}_{\Omega}(u)$ in (5.91) is given by the sum

$$\mathcal{T}(x) = \mathcal{T}_E(x) + \beta \mathcal{T}_V(x) \quad \forall x \in \Omega , \qquad (5.99)$$

where the topological derivative of the volume constraint $\mathcal{T}_V(x)$ is given by (5.92) whereas the topological derivative of the mechanism effectiveness $\mathcal{T}_E(x)$ can be obtained from (5.81), namely

$$\mathcal{T}_E(x) = \mathbb{P}_\gamma \sigma(u(x)) \cdot (\nabla v(x))^s , \qquad (5.100)$$

where u and v are the solutions of (5.96) and (5.98), respectively, and $\mathbb{P}_\gamma$ is the polarization tensor from (5.82).

In order to fix these ideas, let us present a numerical example where the minimization problem (5.91) is solved with the help of Algorithm 1. It consists in an inverter *mechanism design*. The hold-all domain representing the initial guess is given by a square clamped on the left corners, while the loads $q_{in} = (2, 0)$ and $q_{out} = (1, 0)$ are respectively applied on the middle of the left and right edges, respectively. See Fig. 5.10a. The penalty parameter in (5.91) is set as $\beta = 3$ and the weight parameter which appears in (5.97) is given by $\kappa = 10$. Finally, the Young modulus, the Poisson ratio, and the contrast in (5.4) are respectively given by $E = 1$, $v = 0.3$, and $\rho_0 = 10^{-4}$. The amplified deformations of the final obtained solution are presented in Fig. 5.10b, where we observe that the obtained mechanism performs the desired movement. The convergence curves for the angle θ_n and shape functional $J(\Omega_n)$ are shown in Fig. 5.11.

5.4   Final Remarks

In this chapter a topology optimization algorithm based on the topological derivative and the level-set domain representation method has been presented. In particular, Algorithm 1 has been proposed in [11] to achieve a local optimality condition for the minimization problem under consideration, which is given in terms of the topological derivative and an appropriated level-set function. This means that the topological derivative is in fact used within the numerical procedure as a steepest-descent direction similar to methods based on the gradient of the cost functional. The topological derivative represents the *exact* first order variation of the shape

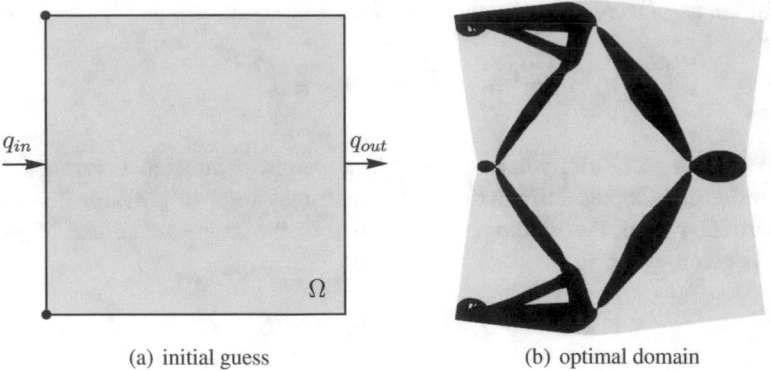

(a) initial guess   (b) optimal domain

Fig. 5.10 Inverter design problem: initial guess and boundary conditions (**a**) and deformed configuration of the optimal domain (**b**)

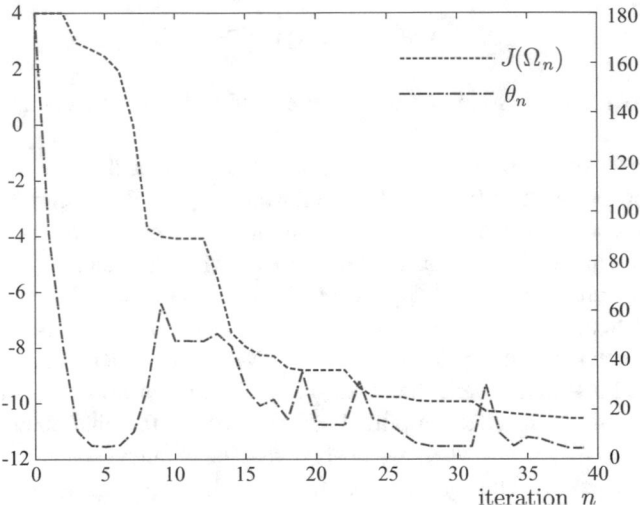

Fig. 5.11 Inverter design problem: convergence curves for the angle θ_n (dashed-dot red line) and shape functional $J(\Omega_n)$ (dashed blue line)

functional with respect to the nucleation of small singular domain perturbations, so that the resulting topology design algorithm converges in few iterations by using a minimal number of user defined algorithmic parameters, as shown in the numerics presented in Sect. 5.3. Furthermore, the topological derivative follows in fact the basic rules of Differential Calculus, which allows for applying it in the context of multi-objective topology optimization algorithms by using e.g., the known formulas already available in the literature. Finally, in contrast to traditional topology optimization methods, the topological derivative formulation does not require any material model concept based on intermediary densities, so that no interpolation schemes are used within the numerical procedures. This feature is crucial in the topology design problem, since the difficulties arising from material model procedures are here naturally avoided. Therefore, the topological derivative method can be seen, when applicable, as a simple alternative method for numerical solution of a wide class of topology optimization problems. For future development of the topological-derivative-based method, we highlight the following:

According to Sect. 5.3, there are numerical evidences showing that Algorithm 1 converges in most cases. However, from the theoretical point of view, only partial results can be found in the literature. See for instance [10], where the convergence of Algorithm 1 has been analyzed in the particular case of an optimal control problem with respect to characteristic functions of small sets. Therefore, the most important theoretical problem to be solved concerns the convergence of Algorithm 1 in general.

The topological derivative concept has also been shown to be effective in solving a certain class of inverse problems [13, 30, 40, 44, 52, 55, 82, 86, 91]. In particular, stability and resolution analysis for a topological-derivative-based imaging functional have been presented in the context of the Helmholtz equation [6]. However, such analysis is missing for other classes of inverse problems, including gravimetry and EIT, for instance. In this direction, a new branch of research arises, which consists in solving a wide range of reconstruction problems with the help of second order topological derivatives [28, 41–43, 53, 66, 84]. In this context, many interesting questions arise, including on how to efficiently use higher order expansions, for instance.

Synthesis and optimal design of materials in a multiscale framework have been considered in [46] and further developed in [16], where the topological derivative of the homogenized elasticity tensor has been obtained. Extension to the dynamic case is a difficult and interesting research topic, where inertial forces acting at the microscale may produce unexpected macroscopic constitutive behavior. Finally, a new emerging research field consists in the design of new materials by considering the strain gradient homogenized constitutive tensor. From the theoretical point of view, a deep question arises in the context of topological derivatives associated with asymptotic models in general, including multiscale and dimension reduction, for instance. In particular, both objects come out from a limit passage procedure, one representing the size of the topological perturbation and the other one controlling the scale. It is not clear whether these limits commute or not. Actually, different results are expected after interchanging the order of these limits.

Topology design of structures taking into account more realistic scenario such as anisotropic elasticity [24, 48], transient wave equations [32], and evolution variational inequalities is a difficult and challenging problem, which requires further development from both theoretical and numerical points of views.

Topological-derivative-based topology design in multiphysics taking into account multiobjective shape functionals is an important and difficult subject of research, which also deserves investigation. Design of antenna and wave guides in nanophotonics is an example of modern application. It can be handled with the use of the domain decomposition technique presented in Chap. 4, for instance.

The Griffith-Francfort-Marigo damage model adopted in [3] and later in [95] and [97] does not distinguish between traction and compression stress states in the damage evolution process. Hence, it is unsuitable for describing the crack closure phenomenon. Therefore, the development of the topological derivative theory for functionals which consider distinct criteria in traction and in compression deserves investigation. However, it is well known that such modeling leads to a class of nonlinear elasticity systems, so that these extensions are expected to be difficult. See also closely related works dealing with crack nucleation sensitivity analysis [8, 94] and crack propagation control [96].

Extension to nonlinear problems in general can be considered as the main challenge in the theoretical development of the topological derivative method. The difficulty arises when the nonlinearity comes out from the main part of the operator, which at the same time suffers a topological perturbation. It is the case of nucleation of holes in plasticity and finite deformations in solid mechanics or small obstacles in compressive fluid flow, for instance. See the recent publication [12] dealing with topological derivatives for a class of quasilinear elliptic equations.

5.5   Exercises

1. By taking $\rho = 1$ in (5.12), derive the Navier system (5.13).
2. From the weak formulation (5.19), derive the strong form (5.20) and discuss the transmission condition on the interface ∂B_ε.
3. By using separation of variable technique, find the stress distribution around the inclusion B_ε, which is the solution of the exterior boundary value problem (5.47).
   Hint: Consult the book by Little [62] and look for the Airy functions in polar coordinate system.
4. Take into account Remark 5.4 and derive the closed formula for the isotropic and uniform fourth order tensor $\mathbb{T}$ given by (5.59) in the form $\mathbb{T} = \alpha_1 \mathbb{I} + \alpha_2 I \otimes I$, by finding the coefficients α_1 and α_2 explicitly.
5. Repeat the derivations presented in Remark 3.1 to find a general representation for the polarization tensor in elasticity.
   Hint: After introducing the notation $w(\varepsilon^{-1}x) := \varepsilon^{-1}w_\varepsilon(x)$ and the change of variable $\xi = \varepsilon^{-1}x$, write $w(\xi)$ as a linear combination of the components of $\sigma(u(\widehat{x}))$ as follows $w(\xi) = \sigma(u(\widehat{x}))_{ij}\, v^{(ij)}(\xi)$. Then replace it into the exterior problem (5.47) to obtain a set of canonical variational problems of the form:

$$v^{(ij)} \in \mathscr{W} : \int_{\mathbb{R}^2} \gamma_\omega \sigma_\xi(v^{(ij)}) \cdot (\nabla_\xi \eta)^s = (1-\gamma)(e_i \otimes e_j) \cdot \int_\omega (\nabla_\xi \eta)^s \quad \forall \eta \in \mathscr{W},$$

$$(5.101)$$

where $\sigma_\xi(v^{(ij)}) = \mathbb{C}(\nabla_\xi v^{(ij)})^s$. The quotient space $\mathscr{W}$ is defined as $\mathscr{W} := \{\varphi \in H^1(\mathbb{R}^2)/\mathbb{R}\}$ and the contrast γ_ω is given by $\gamma_\omega = 1$ in $\mathbb{R}^2 \setminus \omega$ and $\gamma_\omega = \gamma$ in ω. Finally, comeback to (5.76) and write the polarization tensor as follows:

$$\mathbb{P}_\gamma = -(1-\gamma)\left(\mathbb{I} + \frac{1}{|\omega|}\int_\omega \sigma_\xi(v^{(kl)})_{ij}(e_i \otimes e_j \otimes e_k \otimes e_l)\right) . \qquad (5.102)$$

6. Code Algorithm 1 and reproduce the numerical examples presented in Sect. 5.3.
7. Study and discuss the list of open problems presented at the end of Sect. 5.4.

Appendix A
Tensor Calculus

In this appendix some basic results of tensor calculus are recalled, which are useful for the development presented in this monograph. We follow the book by Gurtin [51]. Let us introduce the following notation for $d \geq 2$:

- $a, b, c, d, e \in \mathbb{R}^d$;
- $A, B, C, S, W \in \mathbb{R}^d \times \mathbb{R}^d$;
- φ scalar field;
- u, v vector fields;
- T, U second order tensor fields.

A.1 Inner, Vector, and Tensor Products

The scalar or *inner product* of two vectors a and b is defined as

$$a \cdot b = b^\top a , \tag{A.1}$$

with $\|a\| = (a \cdot a)^{1/2}$ used to denote the *Euclidean norm* of the vector a. The tensor A is a linear map that assigns to each vector a a vector $b = Aa$. The transpose $A^\top$ of a tensor A is the unique tensor with the property

$$a \cdot Ab = A^\top a \cdot b , \tag{A.2}$$

for all vectors a and b. An important tensor is the identity I defined by $Ia = a$ for every vector a. The product of two tensors A and B is a tensor $C = AB$. In general $AB \neq BA$. When $AB = BA$, we say that A and B commute. The scalar or *inner product* of two tensors A and B is defined as

© The Author(s), under exclusive license to Springer Nature Switzerland AG 2020
A. A. Novotny, J. Sokołowski, *An Introduction to the Topological Derivative Method*, SpringerBriefs in Mathematics, https://doi.org/10.1007/978-3-030-36915-6

$$A \cdot B = \mathrm{tr}(B^\top A) = \mathrm{tr}(A^\top B) \quad \Rightarrow \quad \mathrm{tr}(AB) = \mathrm{tr}(BA) , \tag{A.3}$$

where the *trace of a tensor* A is defined as

$$\mathrm{tr}(A) = I \cdot A . \tag{A.4}$$

It then follows that

$$A \cdot (BC) = (B^\top A) \cdot C = (AC^\top) \cdot B . \tag{A.5}$$

The *vector product* of two vectors a and b is defined as

$$a \times b = -b \times a . \tag{A.6}$$

Furthermore

$$a \times a = 0 , \tag{A.7}$$

and

$$a \cdot (b \times c) = c \cdot (a \times b) = b \cdot (c \times a) = \mathrm{vol}(\mathfrak{P}) , \tag{A.8}$$

where $\mathfrak{P}$ is the parallelepiped defined by the vectors a, b, and c. Finally, the determinant of a second order tensor is defined as

$$\det A = \frac{Aa \cdot (Ab \times Ac)}{a \cdot (b \times c)} . \tag{A.9}$$

The *tensor product* of two vectors a and b is a second order tensor $A = a \otimes b$ that assigns to each vector c the vector $(b \cdot c)a$, namely

$$(a \otimes b)c = (b \cdot c)a . \tag{A.10}$$

Then it follows that

$$(a \otimes b)^\top = (b \otimes a) , \tag{A.11}$$

$$(a \otimes b)(c \otimes d) = (b \cdot c)(a \otimes d) , \tag{A.12}$$

$$(a \otimes b) \cdot (c \otimes d) = (a \cdot c)(b \cdot d) , \tag{A.13}$$

$$\mathrm{tr}(a \otimes b) = a \cdot b , \tag{A.14}$$

$$a \cdot Ab = A \cdot (a \otimes b) , \tag{A.15}$$

$$A(a \otimes b) = (Aa) \otimes b . \tag{A.16}$$

A.2 Gradient, Divergence, and Curl

Let us consider the smooth enough fields φ, u, v, T, and U, where φ is scalar, u, v are vectors, and T, U are tensors. Here, we do not state smoothness hypotheses, since standard differentiability assumptions sufficient to make an argument rigorous are generally obvious to mathematicians and of little interest to engineers and physicists. Then the following *tensor calculus identities* hold true:

$$\nabla(\varphi u) = \varphi \nabla u + u \otimes \nabla \varphi \,, \tag{A.17}$$

$$\text{div}(\varphi u) = \varphi \text{div}(u) + \nabla \varphi \cdot u \,, \tag{A.18}$$

$$\text{curl}(\varphi u) = \varphi \text{curl}(u) + \nabla \varphi \times u \,, \tag{A.19}$$

$$\text{curl}\,\text{curl}(u) = \nabla \text{div}(u) - \Delta u \,, \tag{A.20}$$

$$\nabla(u \cdot v) = (\nabla u)^{\top} v + (\nabla v)^{\top} u \,, \tag{A.21}$$

$$\text{div}(u \times v) = u \cdot \text{curl}(v) - v \cdot \text{curl}(u) \,, \tag{A.22}$$

$$\text{div}(u \otimes v) = u\,\text{div}(v) + (\nabla u)v \,, \tag{A.23}$$

$$\text{div}(T^{\top} u) = \text{div}(T) \cdot u + T \cdot \nabla u \,, \tag{A.24}$$

$$\text{div}(\varphi T) = \varphi \text{div}T + T \nabla \varphi \,, \tag{A.25}$$

$$\text{div}(\nabla u^{\top}) = \nabla \text{div}(u) \,, \tag{A.26}$$

$$\text{div}(TU) = (\nabla T)U + T\text{div}(U) \,, \tag{A.27}$$

$$\nabla(T \cdot U) = (\nabla T)^{\top} U + (\nabla U)^{\top} T \,. \tag{A.28}$$

Note that the curl of a vector field u, denoted by $\text{curl}(u)$, is the unique vector field with the following property:

$$(\nabla u - \nabla u^{\top})a = \text{curl}(u) \times a \tag{A.29}$$

for every constant vector a. Therefore, $\text{div}\,\text{curl}(u) = 0$ and

$$\text{curl}(u) = 0 \quad \Leftrightarrow \quad u = \nabla \varphi \,. \tag{A.30}$$

In addition, if

$$\text{div}(u) = 0 \quad \text{and} \quad \text{curl}(u) = 0 \,, \tag{A.31}$$

then u is harmonic, namely $\Delta u = 0$. Finally, the divergence of a tensor field T, denoted as $\text{div}(T)$, is the unique vector field with the following property:

$$\text{div}(T) \cdot a = \text{div}(T^{\top} a) \tag{A.32}$$

for every constant vector a.

A possible representation for the *Dirac mass* is given by a Gaussian distribution. Into two spatial dimensions it can be written as

$$\delta(x - \xi) = \lim_{\varepsilon \to 0} \frac{1}{2\pi\varepsilon^2} \exp\left(-\frac{\|x - \xi\|^2}{2\varepsilon^2}\right).$$ (A.33)

From (A.21), the gradient of $\delta(x - \xi)$ with respect to ξ can be obtained as follows:

$$\nabla_\xi \delta(x - \xi) = \lim_{\varepsilon \to 0} \frac{x - \xi}{2\pi\varepsilon^4} \exp\left(-\frac{\|x - \xi\|^2}{2\varepsilon^2}\right).$$ (A.34)

These results are important in the of monopole and dipole theory, respectively.

A.3   Integral Theorems

Let Ω be an open and bounded domain in $\mathbb{R}^d$, $d \geq 2$, whose boundary is denoted by $\partial\Omega$. Let n denote the outward unit normal vector field on the boundary $\partial\Omega$ of Ω. Here, we state the integral theorems without proofs and without smoothness assumptions regarding the underlying functions and the domain of integration as well. Then, given scalar φ, vector v, and tensor T fields, the following *integral identities* hold true:

$$\int_\Omega \nabla\varphi = \int_{\partial\Omega} \varphi\, n,$$ (A.35)

$$\int_\Omega \nabla v = \int_{\partial\Omega} v \otimes n,$$ (A.36)

$$\int_\Omega \operatorname{div}(v) = \int_{\partial\Omega} v \cdot n,$$ (A.37)

$$\int_\Omega \operatorname{div}(T) = \int_{\partial\Omega} Tn.$$ (A.38)

Divergence theorems are deep mathematical results central to the derivations presented in this monograph. In particular, let us state the *divergence theorems* in their useful forms, namely

$$\int_\Omega (T \cdot \nabla v + \operatorname{div}(T) \cdot v) = \int_\Omega \operatorname{div}(T^\top v) = \int_{\partial\Omega} Tn \cdot v.$$ (A.39)

If $T = S$, with S a symmetric tensor field ($S = S^\top$), then

$$\int_\Omega (S \cdot \nabla v^s + \operatorname{div}(S) \cdot v) = \int_\Omega \operatorname{div}(Sv) = \int_{\partial\Omega} Sn \cdot v.$$ (A.40)

In addition, we have

$$\int_{\Omega} (\varphi \mathrm{div}(v) + \nabla \varphi \cdot v) = \int_{\Omega} \mathrm{div}(\varphi v) = \int_{\partial \Omega} \varphi\, v \cdot n \,. \tag{A.41}$$

If $v = \nabla \phi$, with ϕ a scalar field, then

$$\int_{\Omega} (\varphi \Delta \phi + \nabla \varphi \cdot \nabla \phi) = \int_{\Omega} \mathrm{div}(\varphi \nabla \phi) = \int_{\partial \Omega} \varphi \partial_n \phi \,. \tag{A.42}$$

Finally, an important result which appears in the context of electromagnetism is given by the following integral theorem:

$$\int_{\Omega} (\mathrm{curl}(u) \cdot v - u \cdot \mathrm{curl}(v)) = \int_{\partial \Omega} n \times u \cdot v \,. \tag{A.43}$$

A.4   Some Useful Decompositions

Every tensor A can be *decomposed* uniquely as the sum of a symmetric tensor S and a skew tensor W, namely

$$A = S + W \,, \tag{A.44}$$

where

$$S = \frac{1}{2} \left(A + A^\top \right) \quad \text{and} \quad W = \frac{1}{2} \left(A - A^\top \right) \,. \tag{A.45}$$

We call S the symmetric part of A and W the skew part of A. Therefore, there is a one-to-one correspondence between vectors and skew tensors

$$Wa = w \times a, \quad \text{with} \quad w_1 = W_{32} \,, \quad w_2 = W_{13} \,, \quad w_3 = W_{21} \,, \tag{A.46}$$

where $W = -W^\top$ is a skew or anti-symmetric second order tensor. In addition,

$$[(a \otimes b) - (b \otimes a)]\, c = -[(a \cdot c)b - (b \cdot c)a] = -(a \times b) \times c \,. \tag{A.47}$$

Finally, we have:

- If S is symmetric,

$$S \cdot A = S \cdot A^\top = S \cdot \left[\frac{1}{2}(A + A^\top) \right] \,. \tag{A.48}$$

- If W is skew,

$$W \cdot A = -W \cdot A^\top = W \cdot \left[\frac{1}{2}(A - A^\top) \right] . \qquad (A.49)$$

- If S is symmetric and W is skew,

$$S \cdot W = 0 . \qquad (A.50)$$

- If $A \cdot B = 0$, for every B, then $A = 0$.
- If $A \cdot S = 0$, for every S, then A is skew.
- If $A \cdot W = 0$, for every W, then A is symmetric.

Into two spatial dimensions $\mathbb{R}^2$, a symmetric second order tensor S, namely $S = S^\top$, admits the following spectral *decomposition*

$$S = s_1(e_1 \otimes e_1) + s_2(e_2 \otimes e_2) , \qquad (A.51)$$

where e_1 and e_2 are the eigenvectors of S, whereas s_1 and s_2 are the associated eigenvalues given by

$$s_{1,2} = \frac{1}{2} \left(\mathrm{tr}\, S \pm \sqrt{2 S^D \cdot S^D} \right) , \qquad (A.52)$$

with S^D standing for the deviatory part of the tensor S, that is

$$S^D = S - \frac{1}{2}(\mathrm{tr}\, S)I . \qquad (A.53)$$

Let us consider a two-dimensional open and bounded domain $\Omega \subset \mathbb{R}^2$, whose boundary is denoted by $\partial \Omega$. Let us also introduce two orthonormal vectors n and τ, such that $n \cdot n = 1$, $\tau \cdot \tau = 1$ and $n \cdot \tau = 0$, defined on the boundary $\partial \Omega$, as shown in Fig. A.1. Then, we have that a vector a defined on $\partial \Omega$ can be decomposed as follows:

Fig. A.1 Curvilinear coordinate system on $\partial \Omega$

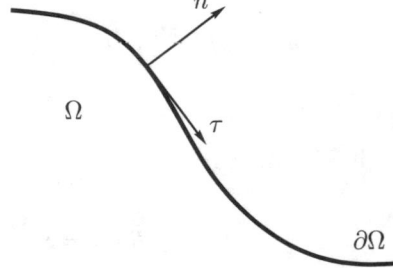

$$a = (n \otimes n)a + (\tau \otimes \tau)a = (a \cdot n)n + (a \cdot \tau)\tau = a^n n + a^\tau \tau , \qquad (A.54)$$

where $a^n := a \cdot n$ and $a^\tau := a \cdot \tau$ are the normal and tangential components of the vector a, respectively. In other words, a^τ is the projection of a into the tangential plane to Ω and a^n is the projection of a orthogonal to the referred tangent plane. In addition, the identity tensor I can be written in the basis (n, τ), namely

$$I = n \otimes n + \tau \otimes \tau . \qquad (A.55)$$

Thus, the projections operators into the tangential and normal directions can respectively be defined as

$$(I - n \otimes n)a = a - (a \cdot n)n = a^\tau \tau , \qquad (A.56)$$

$$(I - \tau \otimes \tau)a = a - (a \cdot \tau)\tau = a^n n . \qquad (A.57)$$

Let A be a second order tensor. Then, A can be decomposed in the basis (n, τ) in the following form:

$$A = A^{nn}(n \otimes n) + A^{n\tau}(n \otimes \tau) + A^{\tau n}(\tau \otimes n) + A^{\tau\tau}(\tau \otimes \tau) , \qquad (A.58)$$

whose components A^{nn}, $A^{n\tau}$, $A^{\tau n}$, and $A^{\tau\tau}$ are defined as

$$\begin{aligned}
An &= \left[A^{nn}(n \otimes n) + A^{n\tau}(n \otimes \tau) + A^{\tau n}(\tau \otimes n) + A^{\tau\tau}(\tau \otimes \tau) \right] n \\
&= A^{nn}(n \cdot n)n + A^{n\tau}(\tau \cdot n)n + A^{\tau n}(n \cdot n)\tau + A^{\tau\tau}(\tau \cdot n)\tau \\
&= A^{nn}n + A^{\tau n}\tau \;\Rightarrow\; A^{nn} = n \cdot An \quad \text{and} \quad A^{\tau n} = \tau \cdot An ,
\end{aligned} \qquad (A.59)$$

$$\begin{aligned}
A\tau &= \left[A^{nn}(n \otimes n) + A^{n\tau}(n \otimes \tau) + A^{\tau n}(\iota \otimes n) + A^{\tau\tau}(\tau \otimes \tau) \right] \tau \\
&= A^{nn}(n \cdot \tau)n + A^{n\tau}(\tau \cdot \tau)n + A^{\tau n}(n \cdot \tau)\tau + A^{\tau\tau}(\tau \cdot \tau)\tau \\
&= A^{n\tau}n + A^{\tau\tau}\tau \;\Rightarrow\; A^{n\tau} = n \cdot A\tau \quad \text{and} \quad A^{\tau\tau} = \tau \cdot A\tau .
\end{aligned} \qquad (A.60)$$

In the same way, we have that the gradient of a scalar field $\nabla\varphi$ defined on $\partial\Omega$ can be decomposed as

$$\begin{aligned}
\nabla\varphi &= (\nabla\varphi \cdot n)n + (\nabla\varphi \cdot \tau)\tau \\
&= (\partial_n\varphi)n + (\partial_\tau\varphi)\tau \;\Rightarrow\; \partial_n\varphi = \nabla\varphi \cdot n \quad \text{and} \quad \partial_\tau\varphi = \nabla\varphi \cdot \tau ,
\end{aligned} \qquad (A.61)$$

where $\partial_n\varphi$ and $\partial_\tau\varphi$ are the normal and tangential derivatives of the scalar field φ. In addition, the gradient of a vector field ∇u defined on $\partial\Omega$ can be decomposed as

$$\nabla u = \partial_n u^n(n \otimes n) + \partial_\tau u^n(n \otimes \tau) + \partial_n u^\tau(\tau \otimes n) + \partial_\tau u^\tau(\tau \otimes \tau) , \qquad (A.62)$$

whose components $\partial_n u^n$, $\partial_\tau u^n$, $\partial_n u^\tau$, and $\partial_\tau u^\tau$ are defined as

$$(\nabla u)n = \left[\partial_n u^n (n \otimes n) + \partial_\tau u^n (n \otimes \tau) + \partial_n u^\tau (\tau \otimes n) + \partial_\tau u^\tau (\tau \otimes \tau) \right] n$$
$$= (\partial_n u^n)n + (\partial_n u^\tau)\tau \ \Rightarrow \ \partial_n u^n = n \cdot (\nabla u)n \quad \text{and} \quad \partial_n u^\tau = \tau \cdot (\nabla u)n,$$
$$(\text{A.63})$$

$$(\nabla u)\tau = \left[\partial_n u^n (n \otimes n) + \partial_\tau u^n (n \otimes \tau) + \partial_n u^\tau (\tau \otimes n) + \partial_\tau u^\tau (\tau \otimes \tau) \right] \tau$$
$$= (\partial_\tau u^n)n + (\partial_\tau u^\tau)\tau \ \Rightarrow \ \partial_\tau u^n = n \cdot (\nabla u)\tau \quad \text{and} \quad \partial_\tau u^\tau = \tau \cdot (\nabla u)\tau.$$
$$(\text{A.64})$$

A.5   Polar and Spherical Coordinate Systems

Let us consider a polar *coordinate system* of the form (r, θ) with center at the origin $\mathcal{O}$, as shown in Fig. A.2. The oriented basis defining this system is denoted by e_r and e_θ, with $e_r \cdot e_\theta = 0$ and $\|e_r\| = \|e_\theta\| = 1$. Thus, we have the representations below.

- Gradient of a scalar field φ:

$$\nabla \varphi = \frac{\partial \varphi}{\partial r} e_r + \frac{1}{r} \frac{\partial \varphi}{\partial \theta} e_\theta . \qquad (\text{A.65})$$

- Laplacian of a scalar field φ:

$$\Delta \varphi = \frac{\partial^2 \varphi}{\partial r^2} + \frac{1}{r} \frac{\partial \varphi}{\partial r} + \frac{1}{r^2} \frac{\partial^2 \varphi}{\partial \theta^2} . \qquad (\text{A.66})$$

- Gradient of a vector field v:

$$\nabla v = \frac{\partial v^r}{\partial r} e_r \otimes e_r + \frac{1}{r} \left(\frac{\partial v^r}{\partial \theta} - v^\theta \right) e_r \otimes e_\theta$$
$$+ \frac{\partial v^\theta}{\partial r} e_\theta \otimes e_r + \frac{1}{r} \left(\frac{\partial v^\theta}{\partial \theta} + v^r \right) e_\theta \otimes e_\theta . \qquad (\text{A.67})$$

- Divergence of a vector field v:

$$\text{div}(v) = \frac{\partial v^r}{\partial r} + \frac{1}{r} \left(\frac{\partial v^\theta}{\partial \theta} + v^r \right) . \qquad (\text{A.68})$$

Fig. A.2 Polar coordinate
system (r, θ)

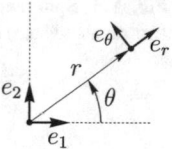

- Divergence of a second order tensor field T:

$$
\text{div}(T) = \left(\frac{\partial T^{rr}}{\partial r} + \frac{1}{r} \frac{\partial T^{\theta r}}{\partial \theta} + \frac{T^{rr} - T^{\theta\theta}}{r} \right) e_r
$$

$$
+ \left(\frac{\partial T^{r\theta}}{\partial r} + \frac{1}{r} \frac{\partial T^{\theta\theta}}{\partial \theta} + \frac{T^{r\theta} + T^{\theta r}}{r} \right) e_\theta \ . \qquad \text{(A.69)}
$$

- Transformation of a vector v from Cartesian to polar:

$$
\begin{pmatrix} v^r \\ v^\theta \end{pmatrix} = \begin{pmatrix} \cos\theta & \sin\theta \\ -\sin\theta & \cos\theta \end{pmatrix} \begin{pmatrix} v^1 \\ v^2 \end{pmatrix} , \qquad \text{(A.70)}
$$

where $v^i = v \cdot e_i$ are the components of vector v in the Cartesian coordinate
system.
- Transformation of a second order tensor T from Cartesian to polar:

$$
\begin{pmatrix} T^{rr} & T^{r\theta} \\ T^{\theta r} & T^{\theta\theta} \end{pmatrix} = \begin{pmatrix} \cos\theta & \sin\theta \\ -\sin\theta & \cos\theta \end{pmatrix}^\top \begin{pmatrix} T^{11} & T^{12} \\ T^{21} & T^{22} \end{pmatrix} \begin{pmatrix} \cos\theta & \sin\theta \\ -\sin\theta & \cos\theta \end{pmatrix} , \qquad \text{(A.71)}
$$

where $T^{ij} = e_i \cdot T e_j$ are the components of tensor T in the Cartesian coordinate
system.

Let us consider a ball $B_\rho(\mathscr{O}) \subset \mathbb{R}^2$ of radius ρ and center at the origin $\mathscr{O}$,
whose boundary is denoted by ∂B_ρ. Then, the integral of a scalar field φ over B_ρ is
evaluated as

$$
\int_{B_\rho} \varphi = \int_0^{2\pi} \left(\int_0^\rho \varphi(r, \theta) r \, dr \right) d\theta \ . \qquad \text{(A.72)}
$$

The integral of φ over the boundary ∂B_ρ is written as

$$
\int_{\partial B_\rho} \varphi = \rho \int_0^{2\pi} \varphi(\rho, \theta) \, d\theta \ . \qquad \text{(A.73)}
$$

Fig. A.3 Spherical
coordinate system (r, θ, ϕ)

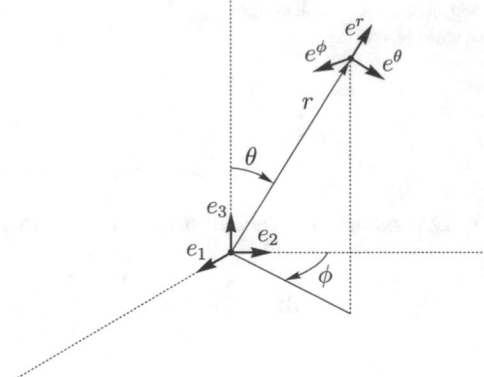

Finally, a general solution for the Laplace equation into two spatial dimensions
can be written in *Fourier series* as follows:

$$\varphi(r, \theta) = A + B \log r + \sum_{k=1}^{\infty} \left[(A_k r^k + B_k r^{-k}) \sin k\theta + (C_k r^k + D_k r^{-k}) \cos k\theta \right].$$
$$(A.74)$$

Let us now consider a spherical *coordinate system* centered at the origin $\mathcal{O}$ given
by (r, θ, ϕ), as shown in Fig. A.3. We define an oriented basis for the system of the
form e_r, e_θ and e_ϕ, with $e_r \cdot e_\theta = e_r \cdot e_\phi = e_\theta \cdot e_\phi = 0$ and $\|e_r\| = \|e_\theta\| = \|e_\phi\| = 1$.
By taking into account this system, we have the representations below.

- Gradient of a scalar field φ:

$$\nabla \varphi = \frac{\partial \varphi}{\partial r} e_r + \frac{1}{r} \frac{\partial \varphi}{\partial \theta} e_\theta + \frac{1}{r \sin \theta} \frac{\partial \varphi}{\partial \phi} e_\phi . \qquad (A.75)$$

- Laplacian of a scalar field φ:

$$\Delta \varphi = \frac{1}{r^2} \frac{\partial}{\partial r} \left(r^2 \frac{\partial \varphi}{\partial r} \right) + \frac{1}{r^2 \sin \theta} \frac{\partial}{\partial \theta} \left(\sin \theta \frac{\partial \varphi}{\partial \theta} \right) + \frac{1}{r^2 \sin^2 \theta} \frac{\partial^2 \varphi}{\partial \phi^2} . \qquad (A.76)$$

- Transformation of a vector v from Cartesian to spherical:

$$\begin{pmatrix} v^r \\ v^\theta \\ v^\phi \end{pmatrix} = \begin{pmatrix} \sin \theta \cos \phi & \sin \theta \sin \phi & \cos \theta \\ \cos \theta \cos \phi & \cos \theta \sin \phi & -\sin \theta \\ -\sin \phi & \cos \phi & 0 \end{pmatrix} \begin{pmatrix} v^1 \\ v^2 \\ v^3 \end{pmatrix}, \qquad (A.77)$$

where $v^i = v \cdot e_i$ are the components of vector v in the Cartesian coordinate
system.

Let us consider a ball $B_\rho(\mathscr{O}) \subset \mathbb{R}^3$ of radius ρ and center at the origin $\mathscr{O}$, with boundary denoted by ∂B_ρ. Then, the integral of a scalar field φ over B_ρ is evaluated as

$$\int_{B_\rho} \varphi = \int_0^{2\pi} \left(\int_0^\pi \left(\int_0^\rho \varphi(r, \theta, \phi) r^2 dr \right) \sin \theta d\theta \right) d\phi . \qquad (A.78)$$

The integral of φ over the boundary ∂B_ρ is given by

$$\int_{\partial B_\rho} \varphi = \rho^2 \int_0^{2\pi} \left(\int_0^\pi \varphi(\rho, \theta, \phi) \sin \theta d\theta \right) d\phi . \qquad (A.79)$$

References

1. G. Allaire, F. de Gournay, F. Jouve, A.M. Toader, Structural optimization using topological and shape sensitivity via a level set method. Control. Cybern. **34**(1), 59–80 (2005)
2. G. Allaire, F. Jouve, H. Maillot, Minimum stress optimal design with the level-set method. Eng. Anal. Bound. Elem. **32**(11), 909–918 (2008)
3. G. Allaire, F. Jouve, N. Van Goethem, Damage and fracture evolution in brittle materials by shape optimization methods. J. Comput. Phys. **230**(12), 5010–5044 (2011)
4. R.C.R. Amigo, S.M. Giusti, A.A. Novotny, E.C.N. Silva, J. Sokolowski, Optimum design of flextensional piezoelectric actuators into two spatial dimensions. SIAM J. Control Optim. **52**(2), 760–789 (2016)
5. H. Ammari, H. Kang, *Polarization and Moment Tensors with Applications to Inverse Problems and Effective Medium Theory*. Applied Mathematical Sciences, vol. 162 (Springer, New York, 2007)
6. H. Ammari, J. Garnier, V. Jugnon, H. Kang, Stability and resolution analysis for a topological derivative based imaging functional. SIAM J. Control Optim. **50**(1), 48–76 (2012)
7. H. Ammari, H. Kang, K. Kim, H. Lee, Strong convergence of the solutions of the linear elasticity and uniformity of asymptotic expansions in the presence of small inclusions. J. Differ. Equ. **254**(12), 4446–4464 (2013)
8. H. Ammari, H. Kang, H. Lee, J. Lim, Boundary perturbations due to the presence of small linear cracks in an elastic body. J. Elast. **113**, 75–91 (2013)
9. S. Amstutz, Sensitivity analysis with respect to a local perturbation of the material property. Asymptot. Anal. **49**(1–2), 87–108 (2006)
10. S. Amstutz, Analysis of a level set method for topology optimization. Optim. Methods Softw. **26**(4–5), 555–573 (2011)
11. S. Amstutz, H. Andrä, A new algorithm for topology optimization using a level-set method. J. Comput. Phys. **216**(2), 573–588 (2006)
12. S. Amstutz, A. Bonnafé, Topological derivatives for a class of quasilinear elliptic equations. J. Math. Pures Appl. **107**, 367–408 (2017)
13. S. Amstutz, N. Dominguez, Topological sensitivity analysis in the context of ultrasonic non-destructive testing. Eng. Anal. Bound. Elem. **32**(11), 936–947 (2008)
14. S. Amstutz, A.A. Novotny, Topological optimization of structures subject to Von Mises stress constraints. Struct. Multidiscip. Optim. **41**(3), 407–420 (2010)
15. S. Amstutz, N. Van Goethem, Topology optimization methods with gradient-free perimeter approximation. Inverse Probl. Imag. **14**(3), 401–430 (2012)

16. S. Amstutz, S.M. Giusti, A.A. Novotny, E.A. de Souza Neto, Topological derivative for multi-scale linear elasticity models applied to the synthesis of microstructures. Int. J. Numer. Methods Eng. **84**, 733–756 (2010)

17. S. Amstutz, A.A. Novotny, E.A. de Souza Neto, Topological derivative-based topology optimization of structures subject to Drucker-Prager stress constraints. Comput. Methods Appl. Mech. Eng. **233–236**, 123–136 (2012)

18. S. Amstutz, A.A. Novotny, N. Van Goethem, Topological sensitivity analysis for elliptic differential operators of order $2m$. J. Differ. Equ. **256**, 1735–1770 (2014)

19. S. Amstutz, C. Dapogny, A. Ferrer, A consistent relaxation of optimal design problems for coupling shape and topological derivatives. Numer. Math. **140**(1), 35–94 (2018)

20. M.P. Bendsøe, Optimal shape design as a material distribution problem. Struct. Optim. **1**(4), 193–202 (1989)

21. M.P. Bendsøe, *Optimization of structural topology, shape, and material* (Springer, Berlin, 1995)

22. M.P. Bendsøe, N. Kikuchi, Generating optimal topologies in structural design using an homogenization method. Comput. Methods Appl. Mech. Eng. **71**(2), 197–224 (1988)

23. D. Bojczuk, Z. Mróz, Topological sensitivity derivative and finite topology modifications: application to optimization of plates in bending. Struct. Multidiscip. Optim. **39**(1), 1–15 (2009)

24. M. Bonnet, G. Delgado, The topological derivative in anisotropic elasticity. Quart. J. Mech. Appl. Math. **66**(4), 557–586 (2013)

25. M. Burger, B. Hackl, W. Ring, Incorporating topological derivatives into level set methods. J. Comput. Phys. **194**(1), 344–362 (2004)

26. R.H. Burns, F.R.E. Crossley, Kinetostatic synthesis of flexible link mechanisms. ASME-Paper 68(36) (1964)

27. D.E. Campeão, S.M. Giusti, A.A. Novotny, Topology design of plates considering different volume control methods. Eng. Comput. **31**(5), 826–842 (2014)

28. A. Canelas, A. Laurain, A.A. Novotny, A new reconstruction method for the inverse source problem from partial boundary measurements. Inverse Probl. **31**(7), 075009 (2015)

29. E.L. Cardoso, J.S.O. Fonseca, Strain energy maximization approach to the design of fully compliant mechanisms using topology optimization. Latin Am. J. Solids Struct. **1**, 263–275 (2004)

30. A. Carpio, M.-L. Rapún, Solving inhomogeneous inverse problems by topological derivative methods. Inverse Probl. **24**(4), 045014 (2008)

31. J. Céa, S. Garreau, Ph. Guillaume, M. Masmoudi, The shape and topological optimizations connection. Comput. Methods Appl. Mech. Eng. **188**(4), 713–726 (2000)

32. I. Chikichev, B.B. Guzina, Generalized topological derivative for the Navier equation and inverse scattering in the time domain. Comput. Methods Appl. Mech. Eng. **194**, 4467–4484 (2008)

33. R. Dautray, J.L. Lions, *Mathematical analysis and numerical methods for science and technology. Volume 2 – Functional and Variational Methods* (Springer, Berlin, 1988)

34. M.C. Delfour, Shape and topological derivatives via one sided differentiation of the minimax of Lagrangian functionals. Int. Ser. Numer. Math. **169**, 227–257 (2018)

35. M.C. Delfour, Topological derivative: a semidifferential via the Minkowski content. J. Convex Anal. **25**(3), 957–982 (2018)

36. H.A. Eschenauer, V.V. Kobelev, A. Schumacher, Bubble method for topology and shape optimization of structures. Struct. Optim. **8**(1), 42–51 (1994)

37. H.A. Eschenauer, N. Olhoff, Topology optimization of continuum structures: a review. Appl. Mech. Rev. **54**(4), 331–390 (2001)

38. J.D. Eshelby, The determination of the elastic field of an ellipsoidal inclusion, and related problems. Proc. R. Soc. A **241**, 376–396 (1957)

39. J.D. Eshelby, The elastic field outside an ellipsoidal inclusion, and related problems. Proc. R. Soc. A **252**, 561–569 (1959)

40. G.R. Feijóo, A new method in inverse scattering based on the topological derivative. Inverse Probl. **20**(6), 1819–1840 (2004)

41. L. Fernandez, A.A. Novotny, R. Prakash, Noniterative reconstruction method for an inverse potential problem modeled by a modified Helmholtz equation. Numer. Funct. Anal. Optim. **39**(9), 937–966 (2018)
42. L. Fernandez, A.A. Novotny, R. Prakash, Topological asymptotic analysis of an optimal control problem modeled by a coupled system. Asymptot. Anal. **109**(1–2), 1–26 (2018)
43. A.D. Ferreira, A.A. Novotny, A new non-iterative reconstruction method for the electrical impedance tomography problem. Inverse Probl. **33**(3), 035005 (2017)
44. J.F. Funes, J.M. Perales, M.L. Rapún, J.M. Manuel Vega, Defect detection from multi-frequency limited data via topological sensitivity. J. Math. Imag. Vision **55**, 19–35 (2016)
45. S.M. Giusti, A.A. Novotny, C. Padra, Topological sensitivity analysis of inclusion in two-dimensional linear elasticity. Eng. Anal. Bound. Elem. **32**(11), 926–935 (2008)
46. S.M. Giusti, A.A. Novotny, E.A. de Souza Neto, R.A. Feijóo, Sensitivity of the macroscopic elasticity tensor to topological microstructural changes. J. Mech. Phys. Solids **57**(3), 555–570 (2009)
47. S.M. Giusti, A.A. Novotny, J. Sokołowski, Topological derivative for steady-state orthotropic heat diffusion problem. Struct. Multidiscip. Optim. **40**(1), 53–64 (2010)
48. S.M. Giusti, A. Ferrer, J. Oliver, Topological sensitivity analysis in heterogeneous anisotropic elasticity problem. Theoretical and computational aspects. Comput. Methods Appl. Mech. Eng. **311**, 134–150 (2016)
49. S.M. Giusti, Z. Mróz, J. Sokolowski, A.A. Novotny, Topology design of thermomechanical actuators. Struct. Multidiscip. Optim. **55**, 1575–1587 (2017)
50. P. Guillaume, K. Sid Idris, The topological asymptotic expansion for the Dirichlet problem. SIAM J. Control Optim. **41**(4), 1042–1072 (2002)
51. M.E. Gurtin, *An introduction to continuum mechanics*. Mathematics in Science and Engineering, vol. 158 (Academic Press, New York, 1981)
52. B.B. Guzina, M. Bonnet, Small-inclusion asymptotic of misfit functionals for inverse problems in acoustics. Inverse Probl. **22**(5), 1761–1785 (2006)
53. M. Hintermüller, A. Laurain, A.A. Novotny, Second-order topological expansion for electrical impedance tomography. Adv. Comput. Math. **36**(2), 235–265 (2012)
54. A.M. Il'in, *Matching of asymptotic expansions of solutions of boundary value problems*. Translations of Mathematical Monographs, vol. 102 (American Mathematical Society, Providence, 1992). Translated from the Russian by V. V. Minachin
55. M. Jleli, B. Samet, G. Vial, Topological sensitivity analysis for the modified Helmholtz equation under an impedance condition on the boundary of a hole. J. Math. Pures Appl. **103**, 557–574 (2015)
56. M. Kachanov, B. Shafiro, I. Tsukrov, *Handbook of elasticity solutions* (Kluwer Academic Publishers, Dordrecht, 2003)
57. V. Kobelev, Bubble-and-grain method and criteria for optimal positioning inhomogeneities in topological optimization. Struct. Multidiscip. Optim. **40**(1–6), 117–135 (2010)
58. V.A. Kozlov, V.G. Maz'ya, A.B. Movchan, *Asymptotic analysis of fields in multi-structures* (Clarendon Press, Oxford, 1999)
59. E. Lee, H.C. Gea, A strain based topology optimization method for compliant mechanism design. Struct. Multidiscip. Optim. **49**, 199–207 (2014)
60. G. Leugering, J. Sokołowski, Topological derivatives for elliptic problems on graphs. Control. Cybern. **37**, 971–998 (2008)
61. G. Leugering, S.A. Nazarov, F. Schury, M. Stingl, The Eshelby theorem and application to the optimization of an elastic patch. SIAM J. Appl. Math. **72**(2), 512–534 (2012)
62. R.W. Little, *Elasticity* (Prentice-Hall, New Jersey, 1973)
63. C.G. Lopes, A.A. Novotny, Topology design of compliant mechanisms with stress constraints based on the topological derivative concept. Struct. Multidiscip. Optim. **54**(4), 737–746 (2016)
64. C.G. Lopes, R.B. Santos, A.A. Novotny, Topological derivative-based topology optimization of structures subject to multiple load-cases. Latin Am. J. Solids Struct. **12**, 834–860 (2015)
65. J. Luo, Z. Luo, S. Chen, L. Tong, M. Yu Wang, A new level set method for systematic design of hinge-free compliant mechanisms. Comput. Methods Appl. Mech. Eng. **198**, 318–331 (2008)

66. T.J. Machado, J.S. Angelo, A.A. Novotny, A new one-shot pointwise source reconstruction method. Math. Methods Appl. Sci. **40**(15), 1367–1381 (2017)

67. W.G. Mazja, S.A. Nasarow, B.A. Plamenewski, *Asymptotics of solutions to elliptic boundary-value problems under a singular perturbation of the domain* (in Russian). (Tbilisi University, Tbilisi, 1981)

68. V.G. Maz'ya, S.A. Nazarov, B.A. Plamenevskij, *Asymptotische theorie elliptischer randwertaufgaben in singulär gestörten gebieten,* vol. 1 (Akademie-Verlag, Berlin, 1991). (English transl.: Asymptotic theory of elliptic boundary value problems in singularly perturbed domains, vol. 1, Basel: Birkhäuser Verlag, 2000)

69. L.R. Meneghelli, E.L. Cardoso, Design of compliant mechanisms with stress constraints using topology optimization. Optim. Struct. Compon. Adv. Struct. Mater. **43**, 35–48 (2013)

70. T. Mura, *Micromechanics of defects in solids* (Kluwer Academic Publishers, Dordrecht, 1987)

71. S.A. Nazarov, Elasticity polarization tensor, surface enthalpy and Eshelby theorem. Probl. Mat. Analiz. **41**, 3–35 (2009). (English transl.: Journal of Math. Sci. 159(1–2):133–167, 2009)

72. S.A. Nazarov, The Eshelby theorem and a problem on an optimal patch. Algebra i Analiz. **21**(5), 155–195 (2009). (English transl.: St. Petersburg Math. 21(5):791–818, 2009)

73. J.A. Norato, M.P. Bendsøe, R.B. Haber, D. Tortorelli, A topological derivative method for topology optimization. Struct. Multidiscip. Optim. **33**(4–5), 375–386 (2007)

74. J.A. Norato, B.K. Bell, D. Tortorelli, A geometry projection method for continuum-based topology optimization with discrete elements. Comput. Methods Appl. Mech. Eng. **293**, 306–327 (2015)

75. A.A. Novotny, J. Sokołowski, *Topological derivatives in shape optimization*. Interaction of Mechanics and Mathematics (Springer, Berlin, 2013)

76. A.A. Novotny, R.A. Feijóo, C. Padra, E. Taroco, Topological sensitivity analysis. Comput. Methods Appl. Mech. Eng. **192**(7–8), 803–829 (2003)

77. A.A. Novotny, R.A. Feijóo, C. Padra, E. Taroco, Topological derivative for linear elastic plate bending problems. Control Cybern. **34**(1), 339–361 (2005)

78. A.A. Novotny, R.A. Feijóo, E. Taroco, C. Padra, Topological sensitivity analysis for three-dimensional linear elasticity problem. Comput. Methods Appl. Mech. Eng. **196**(41–44), 4354–4364 (2007)

79. A.A. Novotny, J. Sokołowski, A. Żochowski, Topological derivatives of shape functionals. Part I: theory in singularly perturbed geometrical domains. J. Optim. Theory Appl. **180**(2), 341–373 (2019)

80. A.A. Novotny, J. Sokołowski, A. Żochowski, Topological derivatives of shape functionals. Part II: first order method and applications. J. Optim. Theory Appl. **180**(3), 683–710 (2019)

81. A.A. Novotny, J. Sokołowski, A. Żochowski, Topological derivatives of shape functionals. Part III: second order method and applications. J. Optim. Theory Appl. **181**(1), 1–22 (2019)

82. J. Rocha de Faria, D. Lesnic, Topological derivative for the inverse conductivity problem: a Bayesian approach. J. Sci. Comput. **63**(1), 256–278 (2015)

83. J. Rocha de Faria, A.A. Novotny, On the second order topological asymptotic expansion. Struct. Multidiscip. Optim. **39**(6), 547–555 (2009)

84. S.S. Rocha, A.A. Novotny, Obstacles reconstruction from partial boundary measurements based on the topological derivative concept. Struct. Multidiscip. Optim. **55**(6), 2131–2141 (2017)

85. L.F.N. Sá, R.C.R. Amigo, A.A. Novotny, E.C.N. Silva, Topological derivatives applied to fluid flow channel design optimization problems. Struct. Multidiscip. Optim. **54**(2), 249–264 (2016)

86. B. Samet, S. Amstutz, M. Masmoudi, The topological asymptotic for the Helmholtz equation. SIAM J. Control Optim. **42**(5), 1523–1544 (2003)

87. O. Sigmund, On the design of compliant mechanisms using topology optimization. Mech. Struct. Mach. Int. J. **25**(4), 493–524 (1997)

88. J. Sokołowski, A. Żochowski, On the topological derivative in shape optimization. SIAM J. Control Optim. **37**(4), 1251–1272 (1999)

89. J. Sokołowski, A. Żochowski, Modelling of topological derivatives for contact problems. Numer. Math. **102**(1), 145–179 (2005)

90. J. Sokołowski, A. Żochowski, Topological derivatives for optimization of plane elasticity contact problems. Eng. Anal. Bound. Elem. **32**(11), 900–908 (2008)
91. R. Tokmashev, A. Tixier, B.B. Guzina, Experimental validation of the topological sensitivity approach to elastic-wave imaging. Inverse Probl. **29**, 125005 (2013)
92. A.J. Torii, A.A. Novotny, R.B. Santos, Robust compliance topology optimization based on the topological derivative concept. Int. J. Numer. Methods Eng. **106**(11), 889–903 (2016)
93. I. Turevsky, S.H. Gopalakrishnan, K. Suresh, An efficient numerical method for computing the topological sensitivity of arbitrary-shaped features in plate bending. Int. J. Numer. Methods Eng. **79**(13), 1683–1702 (2009)
94. N. Van Goethem, A.A. Novotny, Crack nucleation sensitivity analysis. Math. Methods Appl. Sci. **33**(16), 1978–1994 (2010)
95. M. Xavier, E.A. Fancello, J.M.C. Farias, N. Van Goethem, A.A. Novotny, Topological derivative-based fracture modelling in brittle materials: a phenomenological approach. Eng. Fract. Mech. **179**, 13–27 (2017)
96. M. Xavier, A.A. Novotny, J. Sokołowski, Crack growth control based on the topological derivative of the Rice's integral. J. Elast. **134**(2), 175–191 (2018)
97. M. Xavier, A.A. Novotny, N. Van Goethem, A simplified model of fracking based on the topological derivative concept. Int. J. Solids Struct. **139–140**, 211–223 (2018)

Index

Printed in the United States
By Bookmasters